DAMMING THE THREE GORGES

Other Books by Probe International

In the Name of Progress: The Underside of Foreign Aid

Odious Debts: Loose Lending, Corruption, and the
   Third World's Environmental Legacy

DAMMING THE THREE GORGES

What Dam Builders Don't Want You To Know

A Critique of the Three Gorges
Water Control Project Feasibility Study

Second Edition

Edited By Margaret Barber and Gráinne Ryder

Probe International

EARTHSCAN
London • Toronto

First edition published 1990 Probe International
ISBN: 0-919849-10-5

Second edition published 1993 simultaneously in the United
Kingdom by Earthscan Publications Ltd., 120 Pentonville
Road, London N1 9JN; and in Canada by Earthscan Canada,
225 Brunswick Avenue, Toronto, Ontario M5S 2M6.

Copyright © Probe International, 1990, 1993.

All rights reserved

A catalogue record for this book is available from the British
Library.

A catalogue record for this book is available from the National
Library of Canada.

Earthscan Publications Limited is an editorially independent
subsidiary of Kogan Page Limited and publishes in association with the International Institute for Environment and
Development and the World Wide Fund for Nature.

Probe International is a research division of EPRF Energy
Probe Research Foundation.

Cover design by David Sider
Cover artwork, A Grand View of the Three Gorges,
        by Elizabeth Rentzelos
Book design and layout by Elizabeth Rentzelos
Printed by BookCrafters

ISBN Summary for Second Edition
0-919849-18-0 softcover Canada
1 85383 186 7 softcover UK

In Canada, Probe International informs the public about agencies such as the federal government's Canadian International Development Agency and Export Development Corporation, and about international agencies that are funded with Canadian tax dollars, such as the World Bank and other international financial institutions. These national and international agencies have financed the world's worst environmental, social and economic disasters, and they have done so in the name of aid to the Third World. We believe that communities in the Third World, like communities everywhere, should have the right to veto so-called development projects that adversely affect their environments and their lives; we especially object to foisting on Third World peoples pesticides and technologies that are banned in the affluent countries.

In the Third World, we work closely with grass roots organizations that are fighting to protect their environment. We strengthen their hand by publicizing their information worldwide, and by searching out and obtaining for them corporate and aid agency documents that are being kept secret. Likewise, they strengthen our hand with detailed information about how our corporations and aid agencies wreak havoc; they help us to stop destructive projects and to make our institutions accountable.

In a democracy, there is no greater guarantee of justice than the free flow of information. Probe International names names. Because we aren't dependent on governments or industry for our funding, we are free to reveal exactly who is doing what, and when. Probe International is an independent environmental advocacy organization, accountable to the public at large, from whom the bulk of our funding comes in the form of small donations.

Probe International
225 Brunswick Avenue
Toronto, Ontario CANADA
M5S 2M6

Acknowledgements

We wish to give special thanks to Patricia Adams for her rigorous editorial assistance, and her unflagging enthusiasm for the book. Special thanks also to Lawrence Solomon, for his patience and guidance throughout preparation of the book, for his expert help with writing and editing, and for his humour when driving a point home.

We would like to gratefully acknowledge the following people who so generously volunteered their time to help: Norman Houghton, Marcia Ryan, Erica Simmons, and John Thibodeau for editing and proofreading the numerous drafts of each chapter; Dave Hubbel for compiling and proofing all the references; Peter Somers, for his help with the terminology.

For their professional assistance and encouragement, we are grateful to Baruch Boxer, David Dawdy, Chris Elliott, Peter Goodwin, David Melville, Wu Mei, and Zhou Peiyuan.

For their legal advice on the professional responsibilities of Canadian engineers who work on international projects, we wish to express our sincere appreciation to Carolina Gallo, Rick Glofcheski, Harald Mattson, Mark Mattson, and Paul Seitz.

Above all we would like to thank the contributors to this book for their time, expertise, and dedication.

And finally, Probe International would like to thank its 20,000 individual supporters, the Beldon Fund, and the Margaret Laurence Fund for their generous support, which made the publication of this book possible.

This book is dedicated to
the people of the Yangtze River Valley.

Contents

List of Figures and Tables

List of Abbreviations

CIDA   Canadian International Development Agency

CIPM   Canadian International Project Managers Ltd.
                It is made up of private engineering firms
                SNC-Lavalin Inc. and Acres International.

CPPCC   Chinese People's Political Consultative Com-
                mittee

CYJV   CIPM Yangtze Joint Venture. It includes
                CIPM and the international consulting
                wings of two state-owned utilities, Brit-
                ish Columbia Hydro and Hydro-Québec.

MWREP   Ministry of Water Resources and Electric
                Power. It has been divided into the Ministry
                of Water Resources and the Ministry of Energy.

YVPO   Yangtze Valley Planning Office

About the Editors

Gráinne Ryder worked as an engineer in Thailand on village water supply projects for three years before joining Probe International in 1987 as a water resources researcher. She headed an international effort to stop the Three Gorges Project until 1990 when she returned to Thailand to coordinate a campaign against a series of dams on the Mekong River.

Margaret Barber joined Probe International in 1990 with a degree in economics and geography. Her work on the Three Gorges issue has included preparing cases opposing Canadian involvement in the project for submission to the International Water Tribunal in Amsterdam and Canada's engineering associations.

About the Foreword Authors

Dai Qing. A Chinese engineer, environmentalist, and award-winning journalist, Dai Qing was the chief editor of *Yangtze! Yangtze!*, the first Chinese book critical of the Three Gorges Dam. Released in early 1989, the *Far Eastern Economic Review* called *Yangtze! Yangtze!* "a watershed event in post-1949 Chinese politics." For her role in spurring the public debate on the wisdom of the Three Gorges Dam, Dai Qing was arrested and detained without trial in a maximum security prison for ten months. The first edition of *Damming The Three Gorges* was dedicated to Dai Qing.

Niu Kangsheng, M.A. was born in the riverside town of Wanxian on the Yangtze River (which would be drowned by the Three Gorges reservoir). He was Associate Professor of English in Chongqing for 25 years. Now residing in Canada, he is an instructor of Chinese drama and culture at York and Waterloo Universities.

About the Experts

Philip M. Fearnside, Ph.D. is Research Professor at Brazil's National Institute for Research in the Amazon. He has worked in India and travelled extensively in China, including the Three Gorges area. His expertise is in the evaluation of development projects including hydroelectric dams. As a Guggenheim Fellow he studied proposed World Bank projects around the world, including the Three Gorges Project.

Joseph S. Larson, Ph.D. is Professor and Director of the Environment Institute, University of Massachusetts, Amherst, U.S.A. His area of expertise is wetland policy and he has visited the middle and lower Yangtze Valley.

Shiu-hung Luk, Ph.D. is Associate Professor of Geography at the University of Toronto, Canada. As a soil erosion specialist, he is involved with soil conservation research and programs in China.

Vijay Paranjpye, Ph.D. is Professor of Economics at Ness Wadia College of Commerce, India. He is the author of *Evaluating the Tehri Dam* and *High Dams on the Narmada* which evaluate the cost-benefit analyses used to justify India's Tehri Dam and Narmada Valley Project.

Alan Penn, M.Sc. has Master of Science degrees from both Cambridge University, England and McGill University, Canada, and a background in chemical limnology and hydrology. He is an environmental advisor to the Cree Regional Authority, where he has work for 12 years on mercury-related issues in Northern Quebec, Canada.

Vaclav Smil, Ph.D. is Professor of Geography at the University of Manitoba, Canada. He is the author of several books on China's energy and environment including *Energy in China's Modernization.*

David L. Wegner, M.Sc. is an aquatic biologist with a background in engineering. He works with the Glen Canyon Environmental Studies Project of the U.S. Department of Interior, conducting scientific studies to determine the environmental impacts of the Glen Canyon Dam on the Colorado River, U.S.A.

Joseph Whitney, Ph.D. is Chairman and Professor of Geography at the University of Toronto, Canada. With expertise in soil erosion management and environmental impact assessments, he is involved with the design and implementation of soil erosion management projects in China.

Philip B. Williams, Ph.D., P.E. is a hydrologist and engineering consultant, and a partner in Philip B. Williams and Associates, Hydrological Consultants. He is also President of the International Rivers Network, based in San Francisco, U.S.A.

I cannot help shuddering at the time and money that would be demanded by the construction of the Three Gorges Dam, as the Canadian feasibility study indicates: the former is measured by decades and the latter in astronomical figures. For I know, as many people from China do, that in the mountain villages in the Three Gorges area, many people are still trapped in great poverty. The entire belongings of a village couple with several children might amount to nothing more than 70 U.S. dollars. If you care to ask them, "What should come first to make your life better, the Three Gorges Dam or a bag of fertilizer?" I am pretty sure that they will not hesitate to reply, "A bag of fertilizer."

When a fish is trapped in a dry ditch, what would you do to rescue it, if you cherish all good intentions? Give it a bucket of water immediately, or promise it plenty of water from a big river sometime later? Zhuang Zi, a well-known Chinese philosopher, active more than 2,000 years ago, tells us in one of his fables that to save a fish so trapped, you should give it a bucket of water immediately. Many Chinese today, including the mountain villagers, still reason the same way as Zhuang Zi does – down-to-earth and wisely.

I cannot help shuddering at the vastness of the man-made lake that would "rise in the precipitous gorges", for it would submerge one of the cradles of Chinese civilization in deep water and gradually bury it with sand and silt. And when I consider the drastic changes the man-made lake would cause to the environment, I cannot but recall Friedrich Engels' famous remarks to the effect that for every victory Man scores over Nature, Nature will eventually retaliate.

The Three Gorges Dam, if completed, will be the world's largest hydroelectric dam – what glory and what grandiosity! Probably it will be regarded as the Ninth Wonder of the World, closely following the Eighth – Qin Shihuang's Tomb, with its tens of thousands of terra-cotta warriors in formidable battle array. Qin Shihuang, in the four decades of his reign, spared neither labour nor expense to build his splendid tomb, but he never knew that his dynasty would perish only four years after his death.

Niu Kangsheng
Toronto, Canada
September, 1990

Damming The Three Gorges is to have a second edition. It comes at a time when Probe International of Canada has won its case at the International Water Tribunal in Amsterdam against the Three Gorges Dam Project of China and the James Bay Hydroelectric Project of Canada. It comes at a time when the International Coalition Against the Three Gorges Dam has been established. It is also the time when the green movement has received increased understanding and support around the world because the dismantling of communist systems in Eastern Europe and the Soviet Union has led to a deeper understanding of humanity and its relationship with nature.

At the same time, work at the Three Gorges site has begun. Having used all available resources and tricks to have the project officially rubber-stamped by the National People's Congress of China in April, 1992, the determined pro-dam leaders – who continue trying to persuade the opposing and undecided groups within the government – wasted no time in starting the preparatory work for the project. At the dam site, roads are being paved, transmission lines are being set up, and villagers are being moved.

Is the world's largest dam going to be built soon? Can we let these leaders, motivated by self-aggrandisement at the expense of the environment and human lives, get away with it? People who care about China and about the Earth are asking: "What can we do now?"

What can we do? China is still under the control of totalitarian authorities. Decision-making processes in China are conducted in a secretive and unpredictable fashion. Corrupt communist officials from the top rank to the lowest level are now mainly concerned with how to accumulate their own wealth. Ordinary Chinese people are denied a right to air their opinions on public affairs, even when those affairs affect their livelihoods.

Perhaps there is something we can do. We can express our love for the long history and scenic beauty of the Yangtze River which has cradled the Chinese civilization; we can demand that the inherited right to livelihood of the people of the Yangtze – who depend upon and are nurtured by the riches of the river – be restored and respected; we can point out the catastrophic consequences that will result from the abuse of power and the inherent shortcomings of the authoritarian political system; we can tell the people that the natural environment upon which they have depended for generations is under a severe threat, and that those who want to build the megadam are not superior to nature.

We are expressing these thoughts – in our land and to our own people – under great pressure.

However, on the other side of the world, there are people fighting to stop the dam. They are the editors, writers and specialists who composed this book. They are not Chinese, their daily lives will not be directly affected by the building of the Three Gorges Dam. However, they stand up to disclose what is behind the dam. They demonstrate the perilous consequences of unrestrained human ambition; they unveil the lies behind the beautiful promises. They have neither money nor guns. They fight with their wisdom, conscience and sense of responsibility. They demonstrate not the power politics that seems to be the evil force dominating the world, but an admiration of nature and a love for mankind.

Thank you for what you have done for the Yangtze River and China. Thank you on behalf of the people who have been deceived, deprived, and denied a right to appeal.

Dai Qing
Beijing, China
October, 1992

This book is an updated and expanded edition of *Damming The Three Gorges: What Dam Builders Don't Want You To Know*, a critique of a Canadian government-World Bank feasibility study of China's Three Gorges Dam. Originally published in September 1990, this book exposed the flawed analyses and compromised calculations evident in the official justification of a large dam project. Since the first edition was published, others have discovered the same defects in other justifications of other large dam projects.

An independent review of the World Bank-financed Sardar Sarovar Dam on India's Narmada River discovered, as did this book's contributors, that the dam builders failed to employ adequate hydrologic data; that they did not properly consider the backwater effects of the dam, or the downstream impacts of the dam on the people, the estuary and fish stocks; that they exhibited gross delinquency in the handling of environmental matters; and that they failed to prove that the dam would perform as planned. "Assertions," the independent reviewers revealed in their 1992 book entitled *Sardar Sarovar: The Report of the Independent Review*, "have been substituted for analysis."

The importance of these two independent critiques cannot be overstated.

They have exposed a disturbing pattern of omissions, errors, and biases in the official justifications of the dam builders – flaws that have thrived under the cloak of secrecy that shielded them from the light of public scrutiny. Until these two independent critiques were published, the international dam builders – governments, corporations, and international aid agencies – could justify their dams in the name of development and with claims to the national interest, without fear of challenge from a public kept ignorant of their calculations. Now these two critiques have exposed that the dam builders could justify their dams only by denigrating the cultural values of the people affected, by discounting current economic activity in the ecosystems they propose to destroy, by treating the environment as dispensable, by making unscientific and uneconomic choices, and most important, by carelessly or over-confidently assigning risks to others who would not assume those risks for themselves.

The relevance of these two independent critiques go far beyond the Three Gorges and Sardar Sarovar dams, making sense of the bad technical and financial record of large dams around the world, and challenging the wisdom of other large dams, unbuilt but on the drawing boards of the international agencies.

Events have made a second edition of *Damming The Three Gorges* a necessity.

On April 3, 1992, China's National People's Congress gave formal, though not unanimous, approval to the Three Gorges Dam. In its drive for international financing, the Chinese authorities cite the Canadian feasibility study which recommended that the dam would be safe and beneficial. The independent experts who have contributed to this book concluded the opposite: the Canadian feasibility study failed to prove that the Three Gorges Dam was either safe or beneficial.

In the interest of an informed public debate, we reprint this second edition of *Damming The Three Gorges: What Dam Builders Don't Want You To Know*. A new chapter on the problem of sedimentation – which threatens to cripple the dam – has been added, as has a chronicle of the significant political events leading to the approval of the Three Gorges Dam on the mighty Yangtze.

Just weeks before the massacre at Tiananmen Square, China's growing environmental movement had scored a momentous victory by successfully opposing the government's plans to build the massive Three Gorges Dam on the Yangtze River. Vice Premier Yao Yilin had announced that the highly contentious project would be postponed for at least five years, saying that: "people do not need to spend too much energy debating this issue for the time being."[1] The unprecedented public repudiation of the proposed Three Gorges Dam was short-lived, however. It ended at Tiananmen Square, when the critics of the Three Gorges Dam were jailed and silenced along with other members of the pro-democracy movement.

For nearly a decade, citizens' groups outside China had been fighting plans for building the Three Gorges Dam on the Yangtze River. Situated on a spectacular stretch of canyon known as the Three Gorges, this was to be the world's largest hydroelectric dam. Dam builders – the governments, engineering industry, and international aid agencies – believed the Three Gorges Dam would do what no other dam on earth has been designed to do: protect millions of people living along the middle and lower reaches of the river from disastrous floods; generate up to 20,000 megawatts of hydroelectricity for China's energy-hungry industrial centres; and transform a 600-kilometre stretch of the fast-flowing river into a smooth navigable waterway for ocean-going vessels.

To do this would require forcibly relocating up to 1.2 million people, permanently sullying the legendary Three Gorges, drowning up to 32,000 hectares of precious farmland, and disrupting the world's third-largest river – the lifeblood of China's industrial and agricultural heartland.

The wisdom of attempting such a mammoth experiment is apparent primarily to those in the business of building large dams and to those in the business of foreign aid. It certainly is not apparent to the people who bear witness to the destruction dams have brought to their land and rivers. From the Canadian Arctic to the Amazonian rainforest to the vast flood plains of Asia, big dams have flooded fertile river valleys, spread waterborne diseases, destroyed productive fisheries, flooded people's homes and ancestral graves, and stripped whole communities of their cultural heritage and their traditional economies. To compound these indignities, big dams have failed to be the cheap source of renewable energy and boon to economic development that dam builders have promised. U.S. dam fighter Brent Blackwelder writes:

> Big dams and water projects have not only failed to achieve those basic objectives but are also leaving a legacy of unsurpassed cultural destruction, disease, and environmental damage.[2]

But despite the evidence, dam builders steadfastly refuse to grasp that human suffering and environmental destruction is inevitable with large dams. They claim more planning, better management, and more money will ensure better luck next time. Now, in this era of ever larger, ever more risky manipulations of the world's great rivers, the "next time" is the Three Gorges Dam on the Yangtze River. The numbers, the people, and the river may differ from the last, but the creed of the dam builders does not: they claim that the benefits of providing electricity, flood control, and improved navigation on the Yangtze River far outweigh the social and environmental costs; and besides, there is no practical alternative.

Probe International, using Canada's Access to Information Act, obtained the official exposition of this creed as applied to the Three Gorges Dam: a feasibility study, completed in 1989, which recommended that the Three Gorges Dam be built. The $14 million study was financed by the Canadian International Development Agency, supervised by the World Bank, and conducted, in secret, by a Canadian consortium of state-owned utilities and private engineering consulting firms. Upon the study's release, the Canadian government praised the consortium's work as a world-class effort which would help the Chinese government make an informed decision on whether or not to build the dam. Not only would the dam provide flood control, hydro power, and improved navigation, according to the Canadian government, but the project was an opportunity for employment in the impoverished Three Gorges region of China and an opportunity for sustainable development. The plans for relocating up to 1.2 million people were touted as among the best in the world and as for the environment, the study concludes: "The impacts which may occur [would] not affect the overall environmental feasibility [of the project] and may indeed enhance the environment."[3]

Public access to the Canadian feasibility study marks an important precedent for foreign-aid-financed projects. To our knowledge, for the first time anywhere, the general public whose tax dollars have gone to financing the study are able to see the proponents' assessment before a final decision was made and before billions of dollars were committed. To test the validity of the proponents' claims about the Three Gorges Dam, Probe International set about reviewing the 13-volume feasibility study with the help of nine experts from around the world. One of our reviewers, Vaclav Smil, an expert on China's energy and environment, expressed outrage upon seeing the feasibility study, saying, "This is not engineering and science, merely an expert prostitution, paid for by Canadian taxpayers."[4]

The Canadian consortium conducted its feasibility study in cooperation with the same government responsible for the Tiananmen massacre; the Canadian study and a similar Chinese study were carried out simultaneously, in secret. The rest of the world must not forget that the protestors in Tiananmen Square, now silenced, were calling for a more open and democratic process – the very process denied them by the Canadian and Chinese dam builders.

One year after the Tiananmen Square massacre, Premier Li Peng revived discussions about building the Three Gorges Dam. Under the current regime, someone would be imprisoned for producing a critique of a government feasibility study: because they cannot produce such a critique, we must.

Chapter One

Damming the Three Gorges: 1920 – 1993

by Gráinne Ryder and Margaret Barber

On April 3, 1992, China's National People's Congress, China's parliament, erupted in a display of opposition unprecedented for this normally rubber-stamp body. The outburst was the latest in the decades-long dispute over the Three Gorges Dam on China's Yangtze River. Although one-third of the 2600 delegates rejected the project, the dam – touted in China's official media as the biggest public works undertaking since the Great Wall – was approved for inclusion in China's current 10-year plan. It marked the most recent leap forward in the project's turbulent seven-decade-long history.

The Yangtze River, 6300 kilometres long, is the world's third-largest river, with a total drainage area of 1.8 million square kilometres. Springing from the glacial mountains of northern Tibet, the river carves its route through the mountains of southwestern China and then heads northeast to surge through a spectacular 200-kilometre reach of deep, narrow canyon, known as *Sanxia* or Three Gorges. From there, the river widens and meanders across southern China's vast, fertile plains to the East China Sea at Shanghai.

The Yangtze River Valley, encompassing an area roughly one-fifth that of Canada's, is China's agricultural and industrial heartland. Supporting roughly 400 million people, one-third of China's population, the valley produces 40 percent of the nation's grain, 70 percent of its rice, and 40 percent of China's total industrial output.

Although the river and its tributaries are the valley's lifeblood, they have also produced some of China's worst natural flood disasters. Five times this century, Yangtze floodwaters have ravaged the middle and lower valley, killing a total of 300,000 people and leaving millions homeless.

China's attempts to prevent flood disasters while permitting navigation and irrigation date back to ancient times, and have traditionally depended on earthen dykes (or levees).

> From the 5th century B.C. on, Chinese philosophers debated rival theories of river management, mirroring their respective theories of political rule. Taoists believing that rivers should be unconstrained, argued that levees should be low and far apart, allowing the river to seek its own course. Confucians argued for large, high dykes set closely together, tightly controlling the course of the river. This would open up fertile areas along the banks for cultivation, but risked disaster if the levee was breached by floodwaters.[1]

The idea of damming the Three Gorges, first proposed in the 1920s by Dr. Sun Yat-sen, a founder of the People's Republic, marked a departure from traditional Chinese water management, and led to increased reliance on modern hydraulic engineering, particularly large-scale multi-purpose dams.

Protecting roughly 10 million people living along the river banks and flood plains from life-threatening floods is the main rationale for building the Three Gorges Dam. The dam is also intended to generate 15,000 to 20,000 megawatts of hydroelectricity for urban industrial centres, and to improve navigation along the river. The favoured dam plan calls for a height of 185 metres, a reservoir that stretches about 600 kilometres upstream of the Three Gorges, and the forced relocation of about a million people.

Political debate over damming the Three Gorges has been

as tempestuous as the Yangtze River itself. Human ecologist Baruch Boxer writes:

> Planners, ideologues, visionaries and scoundrels alike have used it either to trumpet commitment to nationalistic ideals, assuage national pride, get rich and powerful, or strengthen competing government planning and energy bureaucracies when their autonomy and power were threatened.[2]

But efforts to push the project through China's elaborate decision-making process have been interrupted over the years by war, ideological struggles, the chaos of the Cultural Revolution, economic troubles, and prolonged governmental debate over the project.

In the 1940s, the United States Bureau of Reclamation, America's foremost dam-building agency, cooperated with Chinese engineers to develop preliminary plans for various dam sites. Then in the 1950s, Soviet dam builders worked with the Chinese on preliminary studies until ideological differences soured relations between the two countries.

The Yangtze floods of 1954, which left 30,000 people dead and one million people homeless, brought an unprecedented sense of urgency to damming the Three Gorges. Chairman Mao Tse Tung vowed to speed up preparations for the dam, and the Yangtze Valley Planning Office* was established to conduct specific design and feasibility studies for the Three Gorges Project, as well as to develop an overall plan for water resource development within the entire Yangtze River basin.

In 1958, after lengthy deliberations involving Chairman

---

*The Yangtze Valley Planning Office is now called the Changjiang Water Resources Commission.

Mao, top-level government officials, and Soviet experts, the central government announced that the Three Gorges Dam would be built, but not immediately. Concluding that flood control alone could not justify the dam's construction, they set about redesigning the dam as multi-purpose, for hydropower and navigation as well as for flood control. More studies then evaluated the project in terms of the dam's additional functions.

Since that time, hundreds of government agencies, bureaucracies, and academic bodies have participated in detailed studies on all aspects of the megaproject. But construction plans were delayed for decades because those involved in the deliberations were unable to resolve a number of key issues: the height of the dam and therefore the area to be flooded, which dictates the number of people who would be forced to move; financing the project; the impact of sediment build-up on navigation; reconciling the need for flood control with the dam's other functions; and whether building the dam at all was technically feasible.

China's Dam Builders Gather Experience

While the fate of the Three Gorges was still far from being resolved, Chinese engineers designed and constructed large-scale dams on tributaries of the Yangtze and in the Yellow River Valley. By 1970, the Yangtze Valley Planning Office was ready for the main channel of the Yangtze and construction began on China's largest hydroelectric dam built to date – the Gezhouba Dam. Forty kilometres downstream from the Three Gorges site, this 45-metre-high dam was viewed as the final test before taking on the ultimate challenge of building the gargantuan dam at the Three Gorges.

China's Modernization

In the late 1970s, Premier Deng Xiaopeng's ambitious

modernization plans, which included quadrupling the country's electricity output by the year 2000, boosted enthusiasm for the Three Gorges' tremendous hydropower potential. As planning for the next Five Year Plan (1981 – 1985) got under way, proponents stepped up their efforts to have the project included as a key investment in the plan. Top Chinese leaders appeared eager to proceed but as the Gezhouba Dam neared completion, six years behind schedule and requiring a total investment more than twice that originally estimated,[3] they opted for a more cautious approach and called for more detailed feasibility studies. In 1981, American specialists were invited to visit the Three Gorges site, leading to a five-year agreement with the United States for technical assistance to Chinese dam builders.

Initially, the U.S. experts were highly critical of the project which, at that time, had a proposed dam height of 200 metres and a generating capacity of 25,000 megawatts. They warned that the dam would not solve flooding problems, that it would be a navigational and economic disaster, that it could cause a catastrophe in the valley by triggering landslides or earthquakes, and that it would concentrate too much hydropower capacity in one place, thereby creating a prime target for military attack. Instead, they advocated building a series of smaller dams on Yangtze tributaries. Despite the discouraging reaction from the U.S. experts, proponents within the Ministry of Water Resources and Electric Power, and the Yangtze Valley Planning Office, carried on with their work although they modestly scaled down the project.

China's Environmental Crisis Emerges

In the wake of Deng Xiaopeng's economic reforms and political liberalization, one disturbing truth surfaced which changed the Three Gorges debate forever: China's environment was in a state of crisis. Three decades of unrestrained industrial development and Mao's "grain-first" policy, which promoted the conversion of all available land and forest into grain fields, had caused untold environmental destruction.[4] For the first time, scientists with environmental concerns participated in official deliberations about the Three Gorges Project, and they wasted no time in publishing papers and articles about the dam's potentially destructive affects on agriculture, freshwater and estuarine fisheries, wildlife habitat, and water quality.

The technical problem of keeping the Three Gorges reservoir sediment-free became a broader environmental issue related to upstream land use patterns and environmental conditions. Soil scientists warned that unabated soil erosion in the Yangtze Valley would not only exacerbate flooding but could drastically reduce the Three Gorges reservoir's planned lifetime. Then in 1981, Sichuan province was struck by major floods which both the Ministry of Forestry and Sichuan province attributed to deforestation, not to a lack of reservoir storage capacity, as claimed by the Ministry of Water Resources and Electric Power.

Three Gorges Plans Proceed Amidst Opposition

By 1983, the Yangtze Valley Planning Office had completed a feasibility study recommending that a 175-metre-high dam with a 150-metre reservoir level be built with construction beginning in 1986. In 1984, the State Council (China's cabinet) approved the project "in principle" which meant that the project would be formally adopted as a key investment in the Seventh Five-Year Plan (1986-1990) pending the approval of the National People's Congress at their spring session in 1985.

But before this could happen, the Ministry of Communications and the municipality of Chongqing criticized the 150-metre-high reservoir level and called for a 180-metre reservoir level instead. And so began another round of criticism; but this time academics, intellectuals, and the press voiced their concerns about the dam's environmental consequences, not least among them the sullying of the legendary Three Gorges which fine arts historian, Wu Shenfeng, describes as "the treasure of our motherland."[5]

Opposition from the Chinese People's Political Consultative Committee

Opposition to the Three Gorges Project gained increasing national and international attention when the Chinese People's Political Consultative Committee (CPPCC) conducted a 38-day field trip in 1986 to gather opinions about the dam. Eminent CPPCC members visited eight cities that would be affected by the dam and convened over 40 open forums to hear from all concerned ministries and bureaus, from experts and scholars, and from local and national CPPCC members. At the end of their trip, they submitted their findings to the Central Committee of the Communist Party and the State Council, with the carefully phrased recommendation that the project should not go ahead in the short term.[6] Specifically, the CPPCC report made a number of strong conclusions:

- **Cost:** The main investment required would be over $21 billion, three times the official Chinese estimates.
- **Flood Control:** The project would not solve flooding in the middle and lower reaches of the Yangtze River; it would increase the severity of flood damage in the upper reaches.
- **Sediment:** The proposed strategy for flushing sediment out of the reservoir was found to be unconvincing.
- **Navigation:** The dam would obstruct, not improve, navigation.
- **Power Generation:** The project would require a high investment and a long construction period.

- **Safety:** The dam would pose a huge risk to the Yangtze Valley because the dam and reservoir could trigger landslides and earthquakes.

The CPPCC conclusions generated a heated debate within China, prompting the National People's Congress to call for further deliberations and to exclude the project from the Seventh Five-Year Plan (1986 – 1990) even though, by this time, the Yangtze Valley Planning Office had completed a preliminary design report and an environmental impact statement for the various dam heights under consideration.

U.S. Dam Builders Propose Joint Venture

Meanwhile in the United States, members of the dam-building industry were vying for direct involvement in the Three Gorges Project. A high-powered consortium known as the U.S. Three Gorges Working Group was formed, and included representatives from the U.S. Bureau of Reclamation, the U.S. Army Corps of Engineers, the American Consulting Engineers Council, Guy F. Atkinson Company, Bechtel Civil and Mineral, Inc., Coopers and Lybrand, Merrill Lynch Capital Markets, Morgan Bank, Morrison-Knudsen Inc., and Stone and Webster Engineering Corporation.

In 1985, the working group submitted a proposal to the Ministry of Water Resources and Electric Power, which reviewed technical aspects of the 180-metre and 150-metre schemes, and recommended social and environmental impact studies. It also recommended that a cost-benefit analysis acceptable to potential financiers be conducted and that the dam be built by a joint venture between the Chinese government and the U.S. Three Gorges Working Group with possible funding from the World Bank, the Asian Development Bank, Sweden, Japan, and Canada.

Expounding the political significance of the proposed

venture, the U.S. proposal states:

> The long friendship and continuing coopera-
> tion between China and the United States is
> the foundation for the successful union on this
> project and others to follow. The waters of the
> Yangtze and the Columbia commingle inevi-
> tably in the Pacific. So too would our interests
> combine in carrying out the Three Gorges
> Project.[7]

U.S.-based environmental groups quickly learned of the
plans and mobilized opposition to U.S. involvement, warning
that the Three Gorges Dam would be the most disastrous dam
ever built.

Canadian Dam Builders Offer Assistance
The Americans had competitors hoping to build the Three
Gorges Dam. Since 1985, Canadian government officials have
been busy meeting Chinese officials lobbying for Canadian
involvement in China's power sector development. In 1986,
the Canadian International Development Agency (CIDA), the
government's foreign aid arm, arranged with China's Ministry of Water Resources and Electric Power for CIDA to finance
a feasibility study to be conducted by a Canadian consortium.
The consortium, known as CIPM Yangtze Joint Venture,
included three private companies, Acres International, SNC,
and Lavalin International,* and two state-owned utilities,

---

*In August 1991, Lavalin Inc., unable to service its Cdn.$200
million debt, sold its engineering contracts to a creditor consortium
of eight Canadian and international banks, which sold them to
SNC Group Inc. SNC-Lavalin Inc., now Canada's largest engineering firm, continues to be a partner in Canadian International
Project Managers Ltd.

Hydro-Québec International, and British Columbia Hydro International. A steering committee including represent-atives from CIDA, the World Bank, and China's Ministry of Water Resources and Electric Power, was formed to super-vise the feasibility study. The World Bank also assembled an international panel of experts whose role, according to panel member and U.S. sediment expert John Kennedy, was solely "to evaluate the study and to assure that it met very high standards of international practice for these kinds of studies."[8]

Both CIDA and the Canadian consortium were enthusias-tic at the prospect of conducting a feasibility study for the largest dam in the world. In the words of Peter Haines, then Vice-President of Professional Services Branch for CIDA:

> Our private sector is living up to its potential in the Three Gorges Project in China – the world's biggest-ever power development. By working closely together, with some support from CIDA, a number of our leading consult-ants and utilities have good prospects of win-ning hundreds of millions of dollars worth of business for Canada.[9]

According to Haines, the Canadian engineers had muscled the U.S. Three Gorges Working Group and others out of the competition for the Three Gorges contract:[10]

> Many countries would give their eye teeth for Three Gorges, but we beat them to the punch. They [the Chinese] like our expertise.[11]

According to the study's terms of reference, its purpose was twofold: "to form the basis for securing assistance from international financial institutions" and "to form an input to [the Chinese] government in its decision-making process."[12]*

Chinese Feasibility Study Commissioned

During 1986, the State Council commissioned its own final exhaustive feasibility study to be conducted under the aegis of the powerful State Planning Commission. Madame Qian Zhengying, then Minister of Water Resources and Electric Power and staunch supporter of the Three Gorges Project, appointed a 400-member Three Gorges Project Examination Committee including not only senior government officials and engineers but also members of the Chinese Academy of Science. Madame Qian also appointed a 12-member panel of Chinese experts to review the committee's findings.

Steering Committee Calls for More Studies

In 1988, as the Canadian study neared completion, the steering committee identified serious gaps in the feasibility study and recommended that "complementary studies" on various environment and resettlement issues be conducted. CIDA granted the Chinese government an additional $4 million with which the Canadian consortium would undertake new studies. CIDA also hired various Canadian consultants to monitor the quality and effectiveness of the work undertaken by the consortium.[13] Peter Haines, the CIDA vice-president who by this time had retired to become a private consultant, was also hired to coordinate the "complementary studies" and advise China on Canadian services, material, and equipment

---

*The terms of reference were leaked to Probe International in 1987 after CIDA, acting on the wishes of the Chinese government, refused to release them. To obtain all the other documents related to, and including the *Three Gorges Water Control Project Feasibility Study*, Probe International had to resort to using the Canadian Access to Information law.

to be offered for the implementation of the project.[14]

Canadian Study Completed

Mounting public criticisms of Canada's role in the Three Gorges Project, combined with public pressure to release the taxpayer-financed feasibility study, prompted CIDA and the consortium to hold a news briefing in February 1989. At that time, CIDA announced completion of the *Three Gorges Water Control Project Feasibility Study** and released the summary volume of the study to Probe International. The feasibility study concluded that a 185-metre-high dam with a reservoir level of 160 metres is technically, environmentally, and economically feasible, and recommended that the project go ahead. When asked whether the roughly one million people in the Three Gorges region who would have to move had been consulted, a representative from China's Ministry of Water Resources and Electric Power replied:

> Convinced of the improvement in terms of living standards and economic prosperity that [the project] will bring about, the local people and the local governments are greatly in favour and have no objections to this plan.[15]

---

*At the news briefing, CIDA released only the summary volume of the 13-volume feasibility study, despite Probe International's request for the entire study, and despite the Chinese government's consent to the release of the entire study. Probe International then appealed to the Information Commissioner appointed by the Canadian government. But the engineering consortium, which had the right to withhold information on the grounds of third-party commercial interest, was apparently threatening CIDA with legal action if CIDA released the balance of the study. Behind closed doors, CIDA and the consortium reached an agreement and the remaining volumes (with many sections censored as per the clauses pertaining to third-party commercial interests) were released to Probe International in April 1989.

Shortly thereafter, the Chinese feasibility study was completed, recommending that a 185-metre-high dam with a reservoir level of 175 metres (not 160 metres as recommended by the consortium) be built starting in 1992. Both feasibility studies were submitted to the State Council for approval and the Chinese government was expected to announce a go-ahead in early May.

Fierce Opposition Delays Dam Decision

Meanwhile, the project began to stir fierce opposition from hundreds of citizens' groups around the world who, having seen the tragic legacy of large dams elsewhere in the world, condemned the proponents' secretive and rosy assessment of the project. Critics believed it would be nothing less than a social and environmental disaster.

Within China, prominent citizens, scientists, intellectuals, and artists gathered to condemn the recommendations that the dam be built and to release *Yangtze! Yangtze!*[16] a collection of interviews and essays critical of the Three Gorges Project. Speaking at the press conference organized by the book's editors, author Zhang Jie declared: "We hope the authorities halt this big-name, big-money, low-benefit project that serves as a monument to a handful of people."[17] The critics' press release stated, "for the first time ordinary Chinese people will not keep silent on a weighty economic policy decision. They don't want to see an endless repetition of foolish policies."[18]

At the spring session of the National People's Congress, hundreds of delegates called for the Three Gorges Project to be postponed into the next century, leading China's Vice-Premier Yao Yilin to announce that a decision to build it had been postponed for at least five years. Days later, the Chinese People's Political Consultative Committee denounced the Chinese feasibility study, saying it was not conducted in a scientific or democratic manner, and called for a complete reappraisal of the entire project under the leadership of the National People's Congress.

This show of opposition to the dam, known to be Premier Li Peng's pet project, was unprecedented and was one of many blows dealt to the Chinese leadership in early 1989. According to sinologist Frederic Moritz of Pennsylvania State University, the premier's "loss of face" due to the Three Gorges Project opposition fuelled the student protests for democracy in Tiananmen Square.[19]

Tiananmen Square Massacre Interrupts Canadian Involvement

The Three Gorges controversy was eclipsed in a matter of weeks by the dramatic demonstrations in Tiananmen Square, which were ended so brutally by the Chinese government on June 4, 1989.

In the aftermath of the June 4 events, Dai Qing – editor-in-chief of *Yangtze! Yangtze!* – was arrested, incarcerated in Qincheng, a top-security jail for political prisoners, and told she was on a list of six people to be executed.[20] Later released, Ms Dai stated: "I was arrested because of my work on the Three Gorges."[21]

Several weeks after the Tiananmen massacre, public outrage over the Chinese government's use of force against its own citizens prompted the Canadian government to suspend several of its development assistance projects to China, including most outstanding work on the Three Gorges Project. CIDA's sole exception involved Peter Haines, whose CIDA contract was extended to June 21, 1991, for the "provision of advisory services to China program *[sic]* in relation to the China Power Sector and the Three Gorges Project." On June 14, 1991, Mr. Haines' company, Professional Resources Inc., signed another CIDA contract to continue providing advisory services for China's Ministry of Energy.[22]

Li Peng Revives Discussions

Despite public condemnations by foreign governments and international financiers, in the year following the Tiananmen massacre, high-level delegations quietly made their way back to the world's greatest untapped market. The U.S. maintained China's coveted "most favoured nation" status, the World Bank resumed lending to China, and while Canada's work on the complementary studies for the Three Gorges Project remained suspended, Canada kept its consultants' foot in the door, and its $2 billion line of trade credit to the Chinese government open. As Deng Xiaopeng predicted, it was just a matter of time before "business as usual" with foreigners would be restored.

In June 1990, only one year after the Tiananmen massacre, Premier Li Peng called a meeting of 76 experts to revive deliberations on the fate of the Three Gorges. Reporting on the findings of the committee members, China's official newspaper, the *People's Daily,* said, "Most favoured the project while some had different opinions or raised some questions."[23]

According to the *South China Morning Post*, the dam builders now had a good chance to get the project off the ground, since the dam's critics had been silenced by the government's campaign against "bourgeois liberalisation."[24] The project was alive once more.

Canadian Engineers Accused of Negligence

But criticism of the project was not silenced abroad, and in a precedent-setting move complaints of negligence, incompetence and professional misconduct were laid against the Canadian engineers that completed the feasibility study.[25] The complaints were lodged by Probe International with the provincial engineering associations – that are legally responsible for maintaining high standards among professional engineers – and were based on the findings contained in the first edition of *Damming The Three Gorges* published in 1990 (see Appendix B for details and outcome).

Probe International complained that the engineers failed to live up to standards of practice required of engineers licensed in Canada, citing their failure to make paramount the protection of the lives and property of those who would be affected by the project. To support its complaints, Probe International cited various infractions: the Canadian engineers recommended leaving half a million people to live in the active flood storage area of the reservoir; they failed to recommend compensation for as many as 30 percent of the urban population who are not registered residents; their seismic and dam safety analyses were inadequate; and the impact of sedimentation was underestimated.

But this challenge to the veracity of the proposed project was not about to interfere with Beijing's schedule.

Damn the Approval. Full Steam Ahead

The Chinese government forged ahead with resettlement "pilot projects" in anticipation of securing formal parliamentary approval. By April, 1991, 10,000 people had been relocated. According to Guo Shuyan, Governor of Hubei province, "we are experimenting with moving households...so that when the project is approved we will be able to carry out the relocation process as quickly and as efficiently as possible."[26]

Quick and efficient indeed. By July 1991 – only 4 months later – 40,000 people had been moved. Wang Jiazhu of the Yangtze Valley Planning Office and chief engineer of the project dismissed the need for due process, saying, "it's out of the question to ask people if they want to be moved or to consult with them."[27]

One resident faced with the prospect of losing her family's ancestral home and small grove of orange trees despaired, "if the government decides to build the dam and you don't agree to move, you can't do anything about it. It really doesn't matter what I think; there's no point in resisting."[28]

As the resettlement of over a million people was getting under way, the momentum of Beijing's campaign to gain international funding for its favoured project was picking up steam. The pro-dam lobby marshalled specious arguments to justify its favoured project.

Floods Give Pro-Dam Lobby a Boost

The floods in June and July 1991, which took the lives of nearly 3,000 people, were cited by the Chinese government to the international community as proof that the Three Gorges Dam must be built for flood protection, and soon. News reports around the world repeated the theme.

But ironically, inside China, even pro-dam officials did not seem so sure. Former Minister of Power in China and long-time dam supporter, Qian Zhengying, said that had the Three Gorges Dam been standing, it would have done little or nothing to prevent the floods, which stemmed from tributaries of the Huai and Hai rivers, downstream of the proposed dam site and close to the eastern seaboard.[29]

Opposition Inside China Revitalized

Meanwhile, as the Chinese government billed the Three Gorges Dam to its international audience as the only way to save millions from disastrous floods, a groundswell of opposition was growing inside China. This opposition came from the academic and scientific community and, for the first time since the Tiananmen Square massacre, their viewpoints began appearing in print inside China.

In an open letter circulated in intellectual and political circles in China, scientist and vice-chair of the CPPCC, Qian Weichang, invoked the Gulf War as a reminder that the Three

Gorges Dam would become a major target should China come under air attack.*[30]

Just a few months later, in January 1992, a Chinese journal carried articles by six Chinese experts critical of the Three Gorges Project. In the journal, social scientist Chen Shaoming warned against wasting China's limited financial resources on such a massive project, and scholar Wang Ping added, "we cannot blindly go ahead with megaprojects."[31]

Soon afterward, at a March 1992 meeting of the Jiusan Society, which consists mainly of Chinese scientists and academics, members criticized the Three Gorges Project. According to Professor Chen Mingshao, vice-president of Beijing Industrial University, "there are very many scientific and technical problems that remain unresolved and require further study and investigation before the project can go ahead." Professor Chen even criticized a report by Vice-Premier Zou Jiahua because it said "that the problem of sedimentation has largely been resolved but I can tell you it has not."[32]

According to the Jiusan Society, other problems that remained unresolved included: population relocation, project cost, threat of earthquakes and threat of war.

The problems noted by the Jiusan Society, and others, continued to be debated by the international community.

International Tribunal Condemns Project

In February 1992, the International Water Tribunal in Amsterdam – established to review cases of water mismanagement and disputes – heard a case concerning the Three Gorges Project brought against the government of

---

*On January 28, 1993, the Peruca Dam in Croatia was seriously damaged when at least five explosive devices – strategically placed throughout the dam's structure – were detonated.

China, the government of Canada, and the Canadian firms that completed the CIDA-financed feasibility study.*

After hearing the evidence about the Canadian feasibility study, the Jury ruled that due to "expediency, the very high ecological and socio-economic risks of the megadam have not been adequately assessed by the defendant's feasibility studies."

The Jury further stated in its ruling that development projects like the Three Gorges Dam "often ignore the rights and interests of the very people they purport to protect and serve." Until the rights of the people of the Yangtze Valley are respected, the Jury ruled, the project should be halted.

Such concerns about the economic, social, and environmental consequences of the project, which preoccupied independent jurists, engineers, scientists and scholars around the world, distressed neither governments eager to renew economic relations with China, nor engineers vying for contracts to build the world's biggest dam.

Foreign Involvement

Just as the opposition to the Three Gorges Project was reappearing in the winter of 1991/92 after the silence that followed the Tiananmen Square massacre, so too did foreign involvement in the project increase at that time. In December 1991, the United States Bureau of Reclamation signed a contract with the Chinese government to provide consulting services and technology for the Three Gorges Dam.[33] The contract also pledged assistance from the United States Army Corps of Engineers.

For its part, the Chinese government again proceeded with efforts to secure international assistance. Reports quoted Zhao Chuanshao of the Ministry of Water Resources saying,

---

*The Three Gorges case was brought by Probe International and the Federation for a Democratic China.

"we need foreign experience during construction, and we are interested in securing foreign loans."[34]

Beijing made this announcement just prior to the March 1992 session of the National People's Congress (NPC), at which approval of the Three Gorges Dam was expected to pass uncontested.

Dam Approved

Days prior to the NPC session and in accordance with NPC rules, delegate Huang Shunxing registered his intention to address the Congress on April 3 about the vote on the Three Gorges Project. He also wanted to present to the Congress a petition opposing the dam signed by Chinese students at universities in several foreign countries.

On the day of the vote, however, the chairman refused to allow any discussion before delegates cast their votes.

In a last-ditch effort to postpone the dam, Mr. Huang interrupted the vote by standing and shouting his opposition. Pandemonium reigned for five minutes as the press corps descended on the delegate. When the chairman yelled above the din that no one would be heard from, another delegate, Liu Caipin, shouted: "The NPC has violated its own law." She then pressed the objection button on her computer and stormed out chanting, "I protest, I protest."[35]

One-third of the NPC delegates registered their concern about the project by voting no or abstaining. But in the end, with two-thirds voting in favour, the Three Gorges Dam was included in the 1990 – 2000 ten-year plan.

A report on the vote in the official *New China News Agency* commented "observers note the process from planning the Three Gorges Project in the early 1950s to today's final approval is a reflection of the scientific and democratic approach in China's major policy decision."[36]

Despite the majority yes vote, foreign agencies were still

unsure about casting their lot in with Beijing on the Three Gorges Dam Project.

Aid Agencies Step Back

Just after the NPC approved the Three Gorges Dam, CIDA announced in Canada's Parliament that it would provide no more money for the project. Although the official explanation for withholding CIDA funds was budgetary restrictions, observers put forth the view that CIDA had buckled under widespread criticism over its participation in the project.[37]

At the same time, sources inside the World Bank said the Bank was "walking on eggs" when it came to the Three Gorges Project, and any World Bank money for the dam would be "camouflaged as some kind of social contribution – roads, schools and so on."[38]

But the Chinese government kept up its search for international support for its favourite project.

China Dangles Bait for Foreign Investors

In July 1992, China announced it would open a huge stretch of territory along the Yangtze River to foreign investment, with the port of Shanghai anchoring the region as a "dragon's head" of economic development. Among the 28 cities included in the scheme were Yichang (near the site of the Three Gorges Dam), Wuhan, Nanjing, Hanzhou and Chongqing.

In August 1992, and for the first time since the Communist takeover of China, the Chinese government allowed foreigners into its finance and insurance markets. A joint venture – including Merrill Lynch & Co. of the United States, a Taiwanese investment firm and the Lippo Group of Indonesia – was given special access to the markets on condition it provide financial support for the dam's construction.[39]

A few months later, the People's Construction Bank of

China (PCBC) opened a branch in Yichang to secure financing for the dam. In addition to acting as cashier of state funds for the dam, the bank's mission was to raise overseas funds by participating in international banking, trust and real estate deals.[40]

But despite having opened its door, and its markets, to foreign investors, Beijing was not attracting the money it had anticipated.

Environmentalists Upset Financing. Beijing Changes Tack

In January 1993, Yu Shizhong, deputy director of the Three Gorges technical advisory commission, admitted: "We know that environmentalists are putting pressure on the World Bank, the Asian Development Bank and foreign governments....We may have trouble getting foreign loans."[41]

But hardware was a different story. According to Mr. Yu, foreign companies that won equipment bids could still obtain concessionary export loans from their governments, adding "the Danes, the Germans, the French and the Americans are all trying to sell us equipment."

To make it easier for potential suppliers, Beijing began holding major exhibitions and symposiums on the Three Gorges Project for "all interested organizations...to display their latest techniques and their experience concerning equipment manufacture, construction technologies and technical services."[42]

After more than seventy years of chaotic debate and planning, the Chinese government was determined to see the Three Gorges Dam under construction before the turn of the century.

Chapter Two
What Dam Builders Don't Want You to Know: A Summary

Nine experts were invited by Probe International to review the Canadian dam builders' *Three Gorges Water Control Project Feasibility Study*. In this chapter, the editors summarize their key findings. The nine chapters to follow provide our experts' detailed analysis.

On Resettlement:

The Canadian dam builders conclude that the people who will be flooded off their land can be resettled onto upland areas.

Our experts have found:

1. The only replacement land available in the region is too steep, too elevated, and too poor to farm.

2. The dam builders know that the most fertile land in the valley would be lost to the reservoir, yet they estimate that one hectare of new land could grow the equivalent of one hectare of submerged land.

3. Farmers who now live above the proposed reservoir level but cultivate land that would be submerged would not receive compensation or replacement land.

4. If people cannot be resettled within the Three Gorges region, the Chinese government could deport them to distant areas now predominately populated by minority groups. The dam builders do not mention this scenario as it would conflict with the World Bank's policy on tribal people and jeopardize future World Bank involvement in the project.

5. The dam builders found that the Chinese government overestimated the amount of available land by 50 percent.

The dam builders extol China's new resettlement policy as being among the best in the world.

Our experts have found:

1. The Chinese government has no intention of compensating from 10 to 30 percent of the people who have illegally migrated to urban centres that would be flooded by the reservoir.

2. The Canadian dam builders consulted with representatives of the central regime who do not necessarily reflect the interests of the local people. The dam builders provide no evidence that local people have had any legitimate input to the project.

3. The resettlement and compensation estimates exclude hundreds of thousands of people who would either lose their farmland permanently or be victim to increasingly severe flooding in the vicinity of Chongqing and in tributary valleys upstream of the dam.

4. The resettlement estimates do not account for natural growth and migration rates to the region which could increase the number of people who would have to be resettled by an additional 100,000 people.

5. Our experts also note that 30 to 40 percent of the 10 million people who have been resettled to make way for large dams in China since the 1950s are still impoverished and lack adequate food and clothing.

On Environment:

The dam builders conclude that impacts which may occur downstream of the Three Gorges Project could enhance the environment.

Our experts have found:

1. The Canadian dam builders based their assessment on a previous Chinese government environmental impact statement, which dismissed the project's environmental effects as insignificant. The Chinese impact statement failed to meet China's National Environmental Protection Agency guidelines.

2. The Canadian dam builders' conclusions are inadequate, misleading, and irresponsible because they neglected the inevitable social and environmental disruption which the dam would cause to the following:

- the 75 million people who live along the Yangtze River downstream of the Three Gorges whose subsistence economies are inextricably tied to the ecosystems along the Yangtze and around the downstream lakes and wetlands.
- land use patterns due to resettlement.
- downstream lakes and wetlands which support productive fisheries and provide critical habitat for endangered Asian waterfowl such as the Siberian crane.
- riparian, estuarine and marine fisheries that are already suffering a serious decline in productivity due to pollution, dammed tributaries, and overfishing.
- coastal flooding and erosion of hundreds of kilometres of China's best farmland.
- the municipality of Shanghai's already saline water supplies during the dry season, which could become even more saline when the river's flow is reduced by operations at the dam.
- wildlife such as the Yangtze River dolphin, the Chinese sturgeon, the finless porpoise and the Yangtze alligator, which are already endangered and could become extinct.

The dam builders state that the most significant environmental impact would be impoundment of the river.
Our experts have found:
1. In addition to creating a 600-kilometre lake, impoundment of the river would flood upstream tributaries and valleys, which would effectively increase the reservoir area by 50 percent over that designated by the dam builders. The dam builders neglected to assess the social and environmental impacts of flooding upstream tributaries and their valleys.

The dam builders predict that both natural fish populations and aquacultural production would increase in the reservoir.
Our experts have found:
1.   World experience with large reservoirs is well-documented and shows that aquacultural productivity in reservoirs plummets after an initial boom in fish populations; there is no evidence in the study to suggest that the Three Gorges reservoir would be an exception.
2.   The dam builders base their assumptions about reservoir productivity on outdated data gathered in the 1950s.

On Flood Control:

The dam builders state that flood control is the primary need for the Three Gorges Project.
Our experts have found:
1.   There is confusion and inconsistency in the dam builders' study as to the area that the Three Gorges Dam would actually be able to protect from Yangtze floodwater.
2.   A crude method of analysis was used to determine flooding patterns downstream of the dam.
3.   The dam builders fail to demonstrate that more people would be protected from flooding downstream by operation of the dam, than would be flooded out as the reservoir level rises during major floods.
4.   Flood benefits are calculated by assuming high growth rates in highly flood-prone areas – a practice which is considered conjectural and invalid by some flood protection agencies.
5.   The area within the reservoir, used for flood storage in the event of a flood, would remain populated with nearly one-half million people. These people, who previously were safe from flooding, would face an increased risk of flooding without any compensation. If compensation was provided the project's cost would increase by 20 percent.

The dam builders predict that up to one-half million people living around the reservoir could be flooded off their land once in their lifetime.

Our experts have found:

1. People around the reservoir could be flooded off their land far more frequently than predicted by the dam builders because their flood probabilities are based on historical records that do not account for increased flooding due to massive deforestation in the watershed. Also, the dam builders do not consider sedimentation in the reservoir or changes in reservoir operation to maximize power generation as factors which could increase the frequency of flooding people living within the reservoir.

On Sedimentation:

The dam builders claim that the Three Gorges reservoir storage capacity can be "preserved indefinitely."

Our experts have found:

1. CYJV arrived at this conclusion without verifying YVPO's original sedimentation data, even though CYJV acknowledged that "the quality and quantity of basic field data is of crucial importance to the sediment load investigation."

2. CYJV admits elsewhere in its study that, after 100 years, 50 percent of the reservoir will be filled. Even after 100 years, sediment build-up in the reservoir will continue.

3. CYJV accepted YVPO's estimate that the bed load (sediment larger than 1 mm) conveyed by the Yangtze is 0.05 percent of the total sediment, an amount so small CYJV decided to count it as 0 in its analysis. This estimate is grossly at odds with the Yichang Hydrological Gauging Station's statistics, which indicate bed load is 1.6 percent of total sediment.

4. If CYJV's calculations are wrong, as the evidence suggests, CYJV's prediction of the amount of reservoir storage that can be preserved by flushing sediment through the dam

is greatly exaggerated. This means that the dam would silt up and have to be decommissioned much sooner than planned.

The dam builders claim to have developed a dam design and operation methodology to deal completely and effectively with the sedimentation problem.

Our experts have found:

1. The operation and design proposed by CYJV to minimize sedimentation problems is unproven. The only attempt to control a river with sediment discharges of this magnitude – at the Sanmenxia Dam on China's Yellow River – is widely recognized as a costly failure.

2. CYJV's dam design and operation methodology is based on untenable, unsubstantiated, and flawed assumptions.

3. In particular, CYJV's prediction that 90-95% of the sediment entering the reservoir will be flushed through, and CYJV's recommended method for doing so, exceed the confidence limits of the science of sediment hydraulics and fluvial geomorphology, and defy characteristics of the Yangtze River for the following reasons:

- Because the flow, sediment transport, and channel characteristics of the Yangtze River and Three Gorges reservoir would be complex, highly variable, and three-dimensional, CYJV's prediction that sediment would deposit in the reservoir at an equilibrium slope – allowing the same quantity of sediment entering the reservoir to be flushed out – is highly unreliable.

- By underestimating bed load, CYJV has underestimated the effect that coarser bed load material has on making the equilibrium slope steeper over time. A steeper slope would cause increased flooding upstream, and shoaling of the navigation channel impeding ship traffic to Chongqing.

4. The "reservoir trap efficiency method" CYJV used to calculate reservoir sedimentation is unable to estimate

sedimentation in the active flood storage zone, and underestimates sedimentation in the dead storage zone of the reservoir.

5. CYJV underestimated total sedimentation rates by ignoring the effect of landslides which could fill several cubic kilometres of the reservoir, further impeding the flushing of sediments through the dam.

The dam builders claim that "reservoir sedimentation will not limit the useful life of the project."

Our experts have found:

1. CYJV's analysis has failed to predict realistically the actual performance of the reservoir. Nor has it carried out a systematic sensitivity analysis of the cumulative effects of uncertainties in its predictions.

2. There is a significant risk that sedimentation in the reservoir would:

- cause aggradation of the river bed upstream past Chongqing, flooding hundreds of thousands of people.
- decrease flood storage capacity and thereby increase the risk of flooding for millions of people who have been induced to move into flood-prone areas downstream.
- cause degradation of the river bed downstream for hundreds of kilometres, eroding flood control embankments, undermining bridge crossings and changing the hydrologic regime of the river on which millions of people depend.
- accelerate coastal erosion, including near Shanghai.
- substantially impair the performance of the project in its economic lifetime.

3. If the Three Gorges Project is completed as planned, it is probable that within a few hundred years the reservoir will almost entirely silt up, creating an unprecedented hazard for the millions of people living downstream.

On Alternatives to the Three Gorges Project:

The dam builders imply that the Three Gorges Project must be built to alleviate China's energy crisis.

Our experts have found:

1. China has a shortage of electricity in its urban centres due to grossly inefficient energy use. The electricity services now being met could be delivered using just 60 percent of the country's existing hydroelectric capacity, leaving China with a reserve of the remaining 40 percent of hydroelectric capacity and all of its fossil fuel plants.

2. Rather than build the Three Gorges Project, a more environmentally sound and cost-effective alternative would be energy efficiency improvements and conservation measures through technology innovations and price reforms.

The dam builders state that there is no practical flood protection alternative in the middle reach of the Yangtze River.

Our experts have found:

1. The dam builders themselves conclude that flood protection dykes surrounding refuge centres are justified wherever the annual probability of flooding is more than 1.5 percent, which is the case for several large areas downstream of the Three Gorges.

2. Flood reduction strategies could achieve superior flood benefits through a combination of upgrading critical dykes, modifying overflow areas, providing refuge centres and protective dykes, improved flood-proofing and flood-warning systems, development zoning, and reforestation.

3. If development in highly flood-prone areas were restricted to 40 percent of that predicted for some of the downstream urban areas, such as the city of Wuhan, the flood damage avoided would equal the dam builders' predicted reduction in flood damages with the project.

On Navigation:

The dam builders conclude that the Three Gorges reservoir would improve navigation between Yichang and Chongqing.

Our experts have found:

1. Navigation on the Yangtze River would be seriously disrupted for at least 18 years while the dam was under construction.

2. The increase in shipping traffic predicted to follow the dam's construction is unsupported by any studies.

3. The dam builders acknowledge that sedimentation in the upstream end of the reservoir could obstruct navigation near Chongqing, but fail to assess the feasibility and cost of massive dredging operations, which could be required.

4. The dam builders acknowledge that shipping traffic could be tripled without the dam by a combination of better traffic control procedures, more powerful tug boats, and extended navigation hours. These improvements to navigation could be made without the risk of obstruction due to sedimentation, as is expected with the dam.

On Design and Safety:

The dam builders claim they have completed an adequate design and safety analysis.

Our experts have found:

1. The dam builders provide an inadequate analysis of a reservoir-induced earthquake.

2. The dam builders provide an inadequate analysis of structural stability.

3. The dam builders underestimate the risks of a catastrophic landslide.

4. The dam builders underestimate the risk of spillway failure.

5. The dam builders fail to consider the downstream effects of cofferdam failure.

6.   The dam builders fail to consider the potential dangers of increased sedimentation in the reservoir which would reduce the dam's ability to hold back large floods.

On Economic Aspects

The dam builders state that the purpose of their study is "to establish firmly whether the Three Gorges Project is technically, economically, and financially feasible on a basis acceptable to international financing institutions."

Our experts have found:

1. The dam builders used a 10 percent discount rate instead of the 12 percent rate which the World Bank, a likely financier, has applied in its economic appraisals of other projects. Using 12 percent, the Three Gorges' net benefits decline by 59 percent.

2.   The dam builders used the Chinese government's artificial exchange rate of 3.7 yuan per U.S. dollar, instead of 5 or 6 yuan to the dollar. At 6 yuan per U.S. dollar, construction costs increase by 30 percent.

Our experts also note:

1.   In China, 2-to-5 year delays in construction are normal. Such a delay would make the project totally unattractive economically. Even a delay of a single year would reduce net benefits by 22.5 percent.

2.   Many factors – such as transmission and distribution losses, and uncounted people in need of resettling – were ignored. Counting these costs would likely make the project non-viable even if everything else went well, which rarely happens.

THE THREE GORGES PROJECT

SOURCE: Canadian International Development Agency, Press Kit, 1989.   Figure 1

The CYJV-Recommended Project

Cost	US $10.7 billion
Location	Sandouping Village
	Xiling Gorge (Three Gorges)
Construction Duration	18 years
Reservoir Length	500 to 600 kilometres
Counties Affected	19
Total Pop. of Counties	14 million
Resettlement	727,000 people
Urban Centres	
Submerged - Fuling	80,000 population
- Wanxian	140,000 population
Number of Towns Submerged	104
Cultivated Land Submerged	14,500 hectares

Cultural/Historical Sites Lost
108 sites identified by CYJV (some dating back to 10,000 B.C.)
Some specific examples include:
- two iron Buddhas, Tianfu Temple, Fengdu City, dating back 600 years through the Ming and Qing Dynasties
- ancient plank road of the Han Period (A.D. 64) along the Daning River through the counties of Wuchan and Wuxi

Chapter Three
Resettlement Plans for China's Three Gorges Dam

by Philip M. Fearnside, Ph.D.

The Three Gorges Project would produce the world's largest dam-displaced population (500,000 – 1,200,000 people), even at the lowest reservoir operating level nominally under consideration. Other Chinese dams have forced major re-settlements – for example, the Danjiangkou Dam on the Han River (380,000), and the Sanmenxia Dam on the Yellow River (320,000).[1] Outside China, the governments of Egypt and Sudan displaced 100,000 people to make way for the Aswan High Dam.

The Purpose of the CYJV Water Control Project Feasibility Study

The executive summary of the *Three Gorges Water Control Project Feasibility Study* describes the mandate of the CIPM Yangtze Joint Venture (CYJV), sponsored by the Canadian International Development Agency, as "to provide an impartial technical review to the Government of China, to assist it in reaching a decision, and to form the basis for securing funding from international institutions."[2] From reviewing the report, it appears that the factions within the government of China that commissioned the study had already reached the conclusion that they wanted the dam, and that the report was to satisfy the second objective: convincing international institutions to fund the project. This makes its mandate inaccurate and its description as "impartial" questionable. The CYJV report is remarkable in the way it strains to emphasize positive aspects of the scheme. Most incredibly, CYJV lists resettlement as a benefit: "resettlement construction

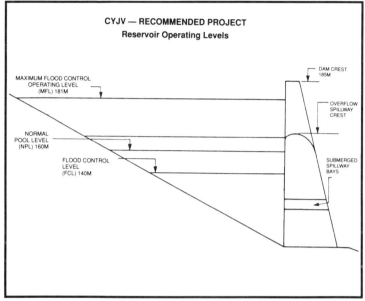

Figure 2

Dam

Dam height (crest)	185 metres (m)
Dam length	2,150 m
Overflow Spillways (26 units)	8 m wide x 20 m high
Submerged Spillways (27 units)	7 m wide x 9 m high

Reservoir

Normal Pool Level* (NPL)	160 m
Flood Control Level (FCL)	140 m
Maximum Flood Control Level (MFL)	181 m

Function

Flood Control Storage	31 x 10^9 cubic metres (m^3)
Installed Hydropower Generating Capacity	16,750 megawatts (MW)
Average Annual Output	68.8 terawatt-hours (TWh)
Navigation Locks	Twin 5-stage flight locks 20 m lift per stage

*Normal Pool Level is the maximum height of the reservoir during the dry season.

and development would spur growth in the area bordering the reservoir."[3] Among the benefits ascribed to the Three Gorges Project is: to "encourage development of the region with resettlement funds."[4] The CYJV steering committee* met six times over the course of the study schedule. At the fourth meeting, the international panel of experts decided that "due to data limitations on the question of land availability for resettlement, resettlement feasibility could not be fully demonstrated."[5] At the sixth and final meeting, the steering committee decided that the terms of reference had been satisfied, and concluded that from the point of view of resettlement the project is feasible – but CYJV does not explain what convinced the panel of experts to reverse their original conclusion. The following discussion refutes the final conclusion reached by CYJV and the panel of experts that resettlement is feasible.

Reservoir Operating Levels

The operating height of the reservoir largely determines the magnitude of impacts due to flooding. The higher the level, the greater the impacts, especially at Chongqing. CYJV recommends 160 metres above sea level as the maximum normal pool level (NPL) because operating levels higher than this would displace many more people: 465,000 if raised to 180 metres. Also, more of the city of Chongqing would be flooded. The report warns: "Neither the economics nor the social impacts of this situation would favour such a reservoir operating condition."[6] CYJV observes that higher operating levels, including a higher flood control level (FCL), would shift

---

*The steering committee is comprised of two representatives from China's Ministry of Water Resources and Electric Power (MWREP), one from the Canadian International Development Agency, and one from the World Bank. The international panel of experts were appointed by MWREP and the World Bank to advise the steering committee.

the fluctuating backwater reach* and its attendant sediment deposition** further upstream, thereby increasing the level of flooding at Chongqing. Because of this, CYJV endorses a normal pool level no higher than 160 metres and a flood control level of 140 metres. However, the Canadian International Development Agency's concluding statement with respect to the study leaves open the possibility of raising operating levels in the future:

> Continuing studies of changing economic factors in China and availability of new data could lead to consideration of slightly modified operating levels.[7]

Expected Flood Levels in the Reservoir

The total number of people to be resettled and compensated is related to the expected level and frequency of floods to be controlled by the dam.*** The CYJV-recommended project, with a normal pool level of 160 metres, would have a dam height of 185 metres. Flood waters would be stored in the reservoir by filling to levels above the flood control level (140 metres) and when the Yangtze's flow exceeds a 50-year flood – the maximum flood expected to occur once every 50 years – floodwaters would be stored above the normal pool level (160 metres). When even heavier floods occur, the reservoir would be allowed to rise up to the maximum flood control operating

---

*The backwater reach of the river is the length of the river upstream of the reservoir whose level fluctuates depending on operating levels of the dam.
**Sediment deposition (or sedimentation) is the process of sediment accumulation in the river. Over time, sediment deposition tends to form a delta in the upstream end of the reservoir which reduces the reservoir's storage capacity and also extends the fluctuating backwater further and further upstream.
***Some people that already live above the designated relocation level may lose farmland to the reservoir that would warrant compensation.

level (MFL) of 181 metres; this level is expected to be reached at a river discharge of 80,000 cubic metres per second (m³/s), an event expected to occur once every thousand years (a 1000-year flood*). These flood probabilities, which are used to set the lower limit for resettlement sites, may well be overly optimistic.

The flood probability analysis is undoubtedly based on the very long series of historical records for the Yangtze River which China is fortunate to have. But the increase in flooding due to deforestation over the past several decades makes simple calculations from the historical record a poor guide to future flood probabilities. For example, a flood almost as "improbable" as the scenario for the maximum operating level at the Three Gorges Project was experienced in 1981, when the flow reached 72,000 m³/s. In all likelihood, the probability of a disastrous flood has increased dramatically in recent years and may continue to increase due to deforestation in the upstream catchment area.[8]

The Yangtze Valley Planning Office (YVPO) designed much of the Chinese government's program for the Three Gorges area. The YVPO plans and the CYJV-recommended project differ in a number of ways, the main difference being that the YVPO plan is less generous in identifying people eligible for compensation and in providing for their livelihoods. It is important to keep both plans in mind, since, now that the project has been approved, events will likely evolve in the direction of the Chinese proposal.

---

*The flood of 1870, with a historical maximum flow of 110,000 m³/s, exceeds the 1000-year flood used by CYJV. The 1870 flood swept away the Zhang Fei Temple (opposite Yunyang) that had been standing since the Three Kingdoms Period (AD 220 – 265). Deforestation since 1870 has undoubtedly increased the danger of major floods such as this one.

YVPO has defined the "relocation and requisition levels"*
along the banks of the Yangtze River using a one-criterion
scheme based on the 20-year flood probability. The river level
corresponding to a 20-year flood is not a horizontal line at a
fixed elevation along the proposed reservoir, but rather slopes
upward from the dam to higher levels in the backwater reach
and tributaries. For most of the reservoir the 20-year flood
level is the same as the normal pool level (160 metres), but at
the upstream end of the reservoir and in the backwater reach
it is higher.

Both YVPO and CYJV have calculated this value using a
Chinese method that considers a water flow of 56,700 m^3/s
rather than using the international method, which would
dictate using 72,000 m^3/s – the flow experienced during the
flood of 1981. Using the former method means that floods can
be expected to occur with greater frequencies than calculations for specific elevations indicate. Therefore more people
deserve compensation than YVPO and CYJV have identified.

CYJV lists YVPO's definition of a requisition level as an
issue that "can be raised with regard to the YVPO methodology."[9] YVPO's requisition levels are based on the 20-year flood
level calculated with the reservoir starting out at the flood
control level (thereby only taking into account flooding in the
summer months). CYJV recommends instead a standard
based on the 20-year flood level with the winter and fall flood
probabilities calculated with the reservoir starting out at the
normal pool level of 160 metres. This would extend the
requisition zone 131 kilometres further upstream, thereby
including more people in the compensation and requisition
estimates.

---

*The Yangtze Valley Planning Office defines the relocation level as
the elevation mark along the reservoir above which resettlement
sites are located. Also, it defines the requisition level as the level
below which various forms of compensation would be given.

The CYJV-recommended project sets "requisition levels" below which compensation would be paid for residences, farmland, factories, and the like. For houses, this level corresponds either to the normal pool level plus two metres[10] or to the 20-year flood level, whichever is higher. People living below these levels would be compensated and moved to the resettlement sites located above 182 metres.

Overall, CYJV praises the YVPO plans. However, closer scrutiny reveals that the plans are not as fair as CYJV leads one to believe. For example, "the loss of the use of land that is flooded in summer but dry in winter (below the natural 20-year flood line) is not included in the compensation criteria."[11] The zone in question is described by CYJV as:

> An unquantified but perhaps substantial area that will be permanently flooded with the reservoir but that is only seasonally or inter-annually flooded under present conditions. This permanent flooding will cause a permanent reduction in the economic activities carried out in this area that will not be compensated.[12]

Not only would the affected farmers be unjustly deprived of land, but the area as a whole would also lose a social function that would not be replaced: this wide strip of land in question is presently farmed during the winter when the river is low and when crops higher up are not producing harvests. It provides wheat, barley, pulses and vegetables for local consumption.

For the people living above the requisition level (those who would remain in the reservoir area), CYJV assumes that the risk of one flood in 20 years is acceptable for farms and houses, and the risk of one flood in 100 years is acceptable for large factories. CYJV claims that compensation would be awarded if flood damage occurs above the requisition levels, but regardless of whether or not compensation would be forthcoming, CYJV is expecting local people to accept a heavy burden of risk. And, should the flood probability calculations prove wrong, the risks could be even more daunting.

CYJV believes that the benefits experienced by the portion of the population who are exposed to flood risk without the dam, and who would be resettled to higher grounds with the dam, counterbalances the added risk of flooding borne by people living between 162 and 182 metres when heavy floods are held back in the reservoir. In short, CYJV concludes that "the trade-off between the before and after situations is more or less even."[13]

YVPO's calculations have resulted in an arbitrarily designated requisition and compensation zone that not only fails to include all of the reservoir but also fails to include the surrounding land that would be subject to an increased risk of flooding. Sedimentation is expected to raise the channel bed in the 120-kilometre reach upstream of the official length of the reservoir. This would enlarge the area and number of people at risk of flooding in future years.[14] And even though YVPO estimates that a 3-metre rise in flood levels at Chongqing would force the relocation of some 90,000 people, it did not account for this in its resettlement estimates.

CYJV estimates the reservoir would extend 498.6 kilometres upstream for a normal pool level of 160 metres. CYJV's estimates of reservoir length result in requisition and compensation zones extended slightly further upstream of the YVPO plans. However, CYJV's zones are, like those defined by YVPO, arbitrarily defined and well downstream of Chongqing which would face increased flooding if the dam is built. For the 160-metre scheme calculated by YVPO, the 5-year flood levels would rise by approximately 7 metres in some places along the reservoir, and by roughly 4 metres near

Chongqing, after 30 years. To calculate this, CYJV inexplicably uses the 5-year flood levels scenario rather than the 20-year flood levels which are used for all other compensation and requisition calculations. Instead of quantifying the magnitude of these potentially enormous impacts to serve as an input for the decision on whether to build the dam, CYJV merely states that:

> Backwater levels affected by reservoir and backwater sedimentation were not included in the [YVPO] assessment. Additional compensation and resettlement resulting from higher sediment-caused backwater levels are deferred to operation of the Three Gorges Project rather than to its construction.[15]

In effect, CYJV has chosen to ignore the additional flooding due to sedimentation.

Furthermore, CYJV failed to quantify either the effects of backwater intrusion or sedimentation along the tributaries in the compensation and resettlement estimates. For the CYJV-recommended scheme of NPL 160 metres, one-third of the reservoir area would consist of flooded tributaries and their valleys. Sedimentation is expected at the confluence of the tributaries (the point where they meet the mainstem of the river) and within the backwater, and could have a particularly significant impact along the Jialing River where it meets the Yangtze River at Chongqing.

In addition to the restricted eligibility for compensation and relocation, and the likelihood that flood levels may be higher in reality than assumed (especially in the backwater reach of the reservoir and in the tributaries), the growth of the population in the reservoir area during construction may be faster than YVPO assumed.* CYJV acknowledges that:

> Higher natural growth and migration rates, and a longer resettlement period [mean that] the CYJV Recommended Scheme population could be increased by an additional 100,000 persons.[16]

Governments Change Their Minds

Many precedents exist where internationally financed development infrastructure has been used for much more damaging ends than originally proposed to financing agencies and addressed in environmental impact studies.[17] The World Bank's internal policies (not any externally mandated restriction) limit the consequences of non-compliance with loan agreement clauses on the environment, resettlement and similar matters, to cutting off the remaining funds for the particular loan in question. Predictably, motivation to comply with such clauses decreases steadily as loan disbursements are received, and disappears completely when disbursements have ended.

In the case of the Three Gorges Project, the Chinese government could promise anything potential financiers might want to hear and then simply change its water management policies once it is no longer constrained by the need to gain approval for international financing. For example, promises could be made to operate the dam at a normal pool level of 160 metres, as recommended by CYJV, and to handsomely compensate all displaced persons. However, once the dam is complete, nothing prevents the Chinese government from changing its mind.

---

*YVPO assumes a 1%/year population growth, a migration rate to the cities of 1.5%/year and to the towns of 0.5%/year.

For example, if funds were to prove insufficient to meet cost overruns, which has traditionally been the case for Chinese dam construction projects such as the Gezhouba Dam, and which cost over twice the amount initially estimated, or even if funds were insufficient to meet the officially approved budget, the Chinese government could decide to discontinue the long-term programs of subsidies and assistance proposed for the reservoir region so that construction of the dam could proceed.*

If the Chinese government wanted to raise the dam's operating levels in order to generate more electricity, the same physical structure could technically be operated with a normal pool level up to 20 metres higher than proposed. Despite the fact that CYJV was unable to confirm the feasibility of resettlement at higher NPLs, one stroke of a pen and one turn of a valve could raise the normal pool level to 180 metres, and as a result, an additional 465,000 people would be displaced with little chance of receiving adequate assistance.

Raising the normal pool level from 160 to 170 metres (and the flood control level from 140 to 145 metres) would generate an additional 800 megawatts of firm energy; however, CYJV states that the value of the power is more than offset by the cost of resettling an additional 260,000 people. Unlike most of the costs of dam building, there are no economies of scale as the number of people displaced increases. CYJV observes that "in fact, experience shows that diseconomies are more likely to occur."[18] CYJV assumes that the additional displaced people would receive the relatively expensive benefits package they have proposed. But the Chinese government could easily raise the water level, thereby cashing in on the power benefits while essentially leaving the people to fend for themselves.

---

*CYJV estimates resettlement costs to be roughly one-third of the total budget for the CYJV-recommended project.

China's own feasibility study, submitted to the State Council on 7 March 1989, recommends a normal pool level of 175 metres.[19]* This level corresponds to a flood control operating level of 145 metres. CYJV warns that operating the dam at a "FCL greater than 145 m [metres] could raise the 100-year flood stage at Chongqing, after about 100 years of Three Gorges Project operation, to around 200 m;"[20] 42 metres higher than the natural water level in Chongqing.

There are a number of strong indications that the Chinese government has plans to fill the reservoir in two stages, going first to the CYJV-recommended normal pool level of 160 metres, followed by filling to either the normal pool level of 175 or 180 metres. A 1987 inundation survey was conducted for a normal pool level (NPL) of 175 metres that "envisages a two stage reservoir filling and related resettlement relocation schedule."[21] Surveys prior to 1987 included even higher NPLs of 180 and 220 metres. Should the Chinese government wish to raise operating levels beyond the limits of the presently proposed structure, it could add height to the dam, as is currently proposed for the Danjiangkou Dam on the Han River (a Yangtze tributary).[22]

"Draft" Plans

Considering plans as eternally in "draft" form is a common practice in many countries, such as Brazil, where the preliminary nature of plans can be used to justify secrecy, and to deflect criticism by alleging that whatever plan is being questioned has changed or is changing. For the Three Gorges

*Editor's Note: A technical description of the Three Gorges Dam, written by the project's chief engineer and an official of China's Ministry of Energy, and published in *Water Power & Dam Construction* in February 1993, confirmed that the normal pool level is to be 175 metres.

Project, "draft" resettlement plans have been under preparation since 1985, with the most recent version prepared in 1987 "to accommodate the Chinese NPL 175 scheme."[23] CYJV states that these plans are "still under review by Chinese authorities and are considered working documents."[24] Such plans could quite easily remain in "draft" status while they are in fact carried out and the dam becomes a concrete and steel reality.

As for construction of the project, China has a tradition of changing plans as construction proceeds, despite efforts to curtail the practice. Major dams have been built using the traditional system of *san bian* (three sides); simultaneously surveying, designing, and building.[25] As examples, the Danjiangkou and Gezhouba dams were both built using this system: both projects had to be halted for two years during construction due to inadequate planning,[26] and their designs were frequently changed throughout the entire construction period.

In the case of the Three Gorges Project, if implementation were to proceed on the recommended scheme of NPL 160 metres, it would be quite a normal practice for plans to evolve toward the more grandiose "draft" proposals for a higher normal pool level or a larger dam. These draft plans would displace more people (especially in Chongqing) and cause even more disruption to local economies and the environment.

The Record of Resettlement

CYJV extols China's new "resettlement with development" policy as "among the best in the world."[27] However, China's record of resettlement is not good, and speaks far louder than a mere announcement of new government policy. Even state planning officials, Tian Fang and Lin Fatang, who tout the virtues of the dam and the possibility of "mobilizing the population" to turn the resettlement areas into regions of prosperity, admit that past resettlements have been plagued with "mistakes such as uncoordinated management, duplicate development, wasteful use of volunteer labour, and limited funds."[28]

In the case of the Danjiangkou Dam, CYJV states that "funds were totally inadequate even for the reconstruction of infrastructure, including housing, at the time of removal. Significant funds for development purposes only became available to relocatees in Yunxian county in 1984, over ten years after resettlement had been completed and even these funds may still be inadequate."[29]

In general, official plans rarely correspond to reality. For example, in the reservoir region, government regulations have prohibited further development below 172 metres, but construction in the doomed cities (below 172 metres) proceeds as usual. CYJV refers to this as "uncontrolled building" and notes that "while the exercise of 'master planning' is taking place, land use planning controls are not very strong."[30]

From this, it is not unreasonable to surmise that the rosy assessment given in the resettlement master plans may never become reality. CYJV acknowledges these doubts in various sections of its study, but these doubts are not reflected in the feasibility study's overall conclusions, which underwrite the Chinese government's plans as adequate.

Land Availability and Quality

In densely populated countries, the most intractable problem associated with dam resettlement schemes has been that land of equivalent quality to the land lost is simply unavailable.* In the case of the Three Gorges Project, the idea

---

*In the case of the Narmada Project in India, which is widely viewed as one of the world's worst dam-building disasters, the lack of replacement land was cited by a World Bank-commissioned independent review, and supported its recommendation that the World Bank pull out of the project until comprehensive resettlement and environment studies could be completed.

that equivalent quality land should be offered to replace the land lost is not even raised as an issue.

The resettlement plans call for relocating people within the same townships (*xiangs*), plus a few additional townships adjacent to those on the reservoir shoreline. Within the 600-kilometre reservoir, many of the displaced rural people would be moved to the eastern end of the region, where YVPO identifies most of the "claimable land." (Claimable land is presently unoccupied land in the zones slated for resettlement that is considered by CYJV to be farmable.)

CYJV's conclusions that "the Chinese plans show that sufficient land resources have been identified to satisfy rural resettlement needs for all affected counties"[31] are doubtful for numerous reasons. With a farmland per capita of 0.067 hectares (less than one-fifth of an acre), there are already too many people farming in the area surrounding the reservoir; 30 percent are considered by the government to be "surplus" due to the contract responsibility system* and agricultural modernization.[32] To increase the amount of "claimable land," CYJV boosted the number of host townships originally proposed by YVPO,** as recommended by the CYJV feasibility study steering committee. But this does not solve the problem for one simple reason: the "claimable land" in the area is unoccupied because the local population has found it to be not worth farming – the soil is poor and the land is too steep to cultivate.

Incredibly, CYJV's analysis of land availability assumes that a hectare of inundated land can be replaced by a hectare of land elsewhere: "preliminary screening of land replacement

---

*The contract responsibility system was instituted in 1981 and motivated an increase in agricultural output by allowing farmers to market some of their produce.
**Host township is the term proponents use for those townships that would not be flooded by the reservoir but would receive an influx of displaced people.

potential was conducted assuming a 1:1 ratio of available to inundated land."[33] This "preliminary screening" was the only one done and is the basis for the CYJV conclusion that sufficient "claimable" land is available. In other words, replacement land is considered to be the same quality as the inundated land – a wholly indefensible premise. In addition to this, CYJV's acceptance of the Chinese expectations for land is unwarranted because:

- The Chinese government has not clearly defined basic characteristics such as altitude, slope and soil type which are necessary for identifying the agricultural potential of unused land.
- CYJV examined the "claimable" land within the 185 to 800 metre elevation band and found only 68 percent as much available land as the Chinese government had asserted.

Later on in the study, CYJV concludes that "taking into account qualitative factors that might increase or decrease the eventual claimable land resource, the [Chinese government's] Three Gorges Project survey results are seen to provide a conservative indication of resource potential."[34] CYJV does not specify what qualitative factors led to its remarkable conclusion. This author believes the quantitative results on land availability and quality support the opposite conclusion, namely that the Chinese government's expectations are overly optimistic.

CYJV only examined land below 800 metres elevation, but the Chinese proposals include plans for resettlement above this mark where over 50 percent of the total "claimable" land is located. Because the number of people to be resettled may be substantially more than officially expected, the Chinese are likely to turn to land above the 800-metre limit when resettlement is actually under way. Unfortunately for the settlers assigned to this zone, the agricultural viability CYJV foresees for lower elevations would not apply.

According to CYJV, apples, pears, apricots, plums and potentially mandarins can be grown, and animal husbandry practised, up to 800 metres elevation. Citrus orchards, except for mandarins, are planned for below 600 metres elevation. Above 800 metres, agriculture is restricted to potatoes, some cereals, and mulberries for silkworm production.

For each crop, the land is considered to be suitable up to a fixed cutoff elevation, and for most citrus varieties this cutoff is at 600 metres because of risk of frost. However, climatic suitability does not, in reality, occur in an all-or-nothing fashion at a given elevation. Rather, production (or frost risk) becomes gradually worse over a wide range of elevations, including the zones defined as suitable. Because land availability increases with elevation, a disproportionate share of the land within each crop's defined zone of suitability would be located near the upper limit of the crop's elevation range. Because of this, the mean crop yields used in the CYJV calculations are likely to be overly optimistic for the land that would actually be planted by resettled farmers within each zone.

Citrus orchards are described by CYJV as the major component in Chinese agricultural plans for the resettlement areas. Plans envisage both intensifying production (by planting citrus) on already cultivated land and bringing uncultivated land into production. But since citrus cannot be grown within the top 200 metres of the elevation range considered by CYJV, as well as on all land above this zone, a large fraction of the farmers would not benefit from the economic potential of the citrus plans.

At the lowest elevations considered for citrus the risk of flooding is the key factor. Like risk of frost at the upper elevations, this can be expected to have a severe impact on the orchards. CYJV considers flooding acceptable for orchards if it does not occur with a frequency greater than once every five years.[35] Under such a flooding regime, the citrus orchards would have to regrow at an unusual pace indeed if they were to remain profitable!

Within the 400 to 800-metre elevation range, the available parcels of land get smaller and smaller as the land improves at lower elevations. The average for the entire elevation range considered by CYJV (up to 800 metres) is a minuscule one hectare (two and a half acres). To farm these plots, villages would have to be divided and dispersed, which would essentially pulverize the existing social structure of villages.* Given this situation, it is difficult to imagine how CYJV intends to satisfy its 14 criteria (Appendix A), one of which states that "opportunities for resettling people in groups and communities should be provided."[36]

Minority Areas

In the semi-arid western regions near China's border with the Soviet Union, in Tibet, or in the tropical areas near the Burmese and Lao borders, settlement projects have been under way as part of the central government's effort to populate the areas with Hans, China's racial and cultural majority.[37] If the plans to reclaim the steep and relatively high-altitude land that makes up most of the claimable category, and to intensify production on already cultivated land, were to prove less successful or more expensive than expected, the temptation would be strong to look for alternative relocation sites further from the Three Gorges region.

---

*In India, where villagers threatened by displacement are permitted to express their views, one of their first demands is to prevent fragmentation of village units. This continues to be a major focus of resistance to the Narmada dams in India.

In this case, there is a possibility that displaced farmers (all of whom belong to the Han majority) could be deported to distant settlement areas presently populated by minority groups. CYJV does not mention this in its report, which is not surprising, because the expressed purpose of the report is to justify the project to international funding agencies. If CYJV were to mention such a scheme as a possibility, it would conflict with the World Bank's policy on tribal peoples[38] and jeopardize future World Bank involvement.

Nonpersons

CYJV fails to mention that the Chinese government apparently has no intention of providing resettlement benefits to "nonpersons." Of urban dwellers, 10 to 30 percent are illegal migrants to the cities (approximately 27,000 to 80,600 people) and therefore are "not officially registered as resident in an urban area."[39] The YVPO surveys of urban areas conducted to estimate resettlement costs did not include these people. CYJV includes this "floating population" in its estimates but uses the most conservative approximation of 10 percent of urban dwellers rather than the more logical midpoint of the range.

The fate of these nonpersons has been a point of contention in China's negotiations with the World Bank. The Chinese government has been averse to providing benefits to these people because it would reward illegal behaviour.[40] Recently, however, the government has softened its position somewhat. The Chinese panel of experts (under China's Leading Group for Three Gorges Project Studies), together with YVPO, now define the affected population as including "the portion of the floating population that has resided in county seats and towns for more than a year."[41] No housing is planned for these people, and, clearly, some of the "floating population" is still excluded from the plans. In addition to the illegal migrants, the government does not officially recognize some whole towns as urban areas, meaning that the residents have no rights to relocation or to industrial jobs.

CYJV appears to endorse the Chinese government's policy of treating many migrants as nonpersons: "CYJV assumed that migrants would not be entitled to resettlement compensation and hence would be discouraged from the zone to be resettled" after resettlement begins.[42] Given that the Chinese government would prefer to provide no resettlement benefits to the floating population, and given that these nonpersons do not even officially exist, it would seem appropriate that some mechanism be identified to ensure that the true size and identity of the floating population is determined, and that these people would not simply be ignored in the resettlement plan.

Rural to Urban Migration Policy

Resettling farmers is much more difficult than moving city people because farming requires land – a finite resource that is fully utilized in the Three Gorges region, and cannot simply be created anew regardless of the amount of funds devoted to that purpose. Therefore, if rural resettlement were to prove unsuccessful, the government has the option of resettling more people in urban areas than originally planned.

In the 1950s and 1960s, millions of people were forcibly relocated from the cities to the countryside. Since that time, government restrictions on internal migration, with people legally tied to their work unit (*dan wei*), have prevented many rural people from responding to the lure of bright city lights. Were these restrictions lifted, some of the "surplus" rural population in the Three Gorges region might willingly move into cities if given the chance.

However, the problem with expanding urban resettlement is that creating factory jobs is about two to three times more expensive per capita than rural resettlement.* Therefore, to allow a greater share of urban resettlement, the budget allocation to resettlement would have to be substantially augmented. But CYJV recommends the opposite: that "measures will have to be adopted and enforced to prevent migration to urban areas below the requisition line."[43] These measures seek to prevent migration within the project area from the countryside to the cities, where new arrivals to the project area would acquire rights to resettlement with urban jobs.

Development Proposals for the Reservoir Region

The Chinese "resettlement with development" policy, as stated by CYJV, includes development assistance to the resettlement areas and favourable prices for power (not necessarily from the Three Gorges Project) to factories in the area. Official policy[44] declares that the Three Gorges region would receive priority for both mining and industrial investments. However, CYJV states that "all large-scale mining projects are controlled entirely by the State, therefore, future development must take into account alternative investments elsewhere in the country. In this regard, the reservoir region is unlikely to receive special attention in the foreseeable future unless special efforts are made to promote these resources."[45] Apparently, the authors of the CYJV study are hinting that official rhetoric on diverting investments from more promising sites is not likely to be translated into tangible benefits for the displaced population. This suspicion is reinforced by a caveat in the official policy decision of the

---

*For example, for the NPL 160 metre scenario, CYJV calculates that agricultural jobs on new land will cost 6,491 yuan/job while jobs in new factories will cost 16,400 yuan/job.

Central Committee of the Communist Party that identified Three Gorges as a special priority:

> Normal expenditures from other channels should not decrease due to the occurrence of resettlement investment. Instead, funds should be raised from all possible means to bring into being a set of industrial and mining enterprises.[46]

Consultation of the Local Population

CYJV refers to public consultation in various sections of the feasibility study. For example, one of CYJV's 14 criteria for "successful resettlement" (Appendix A) states that:

> The resettlement plans should have broad-based popular acceptance and the affected population should be consulted.[47]

In addition, the international panel of experts lists as a "factor to be considered" that:

> It is especially important for the success of the Three Gorges Project resettlement program that not only the population to be resettled but also the host population are convinced that they will benefit from the Project.[48]

CYJV provides no evidence that the local populations have had any legitimate input to the project planning and decision making, much less that they are "convinced" of the benefits. Moreover, CYJV's statement that "the expectations and concerns of the host population are not so well known"[49] suggests that the host population may not have been consulted at all.

The reservoir area as surveyed is made up of 610 townships (*xiangs*). Approximately 500 townships, all but a few of which are in the surveyed area, would receive displaced farmers according to plans for the CYJV-recommended scheme. The CYJV team visited China in March and April 1988 for an unspecified length of time, during which 43 townships (*xiangs*), a mere 10 percent of the townships which would receive displaced farmers, were subjected to "an independent check, using aerial photo interpretation techniques supported by field verification."[50] The study does not indicate how many of the 43 townships they visited nor how much time was devoted to their visits. This 10 percent sample is described by CYJV as "representing typical conditions throughout the reservoir area."[51] While this may be true for physical factors, which can be checked from maps and aerial photos, these methods cannot guarantee representativeness from the standpoint of local consultation and support for the project. Almost universally, governments requesting international financing for projects tend to show visiting delegations carefully selected "success stories." Nothing is said of what steps might have been taken to avoid this problem.

CYJV states that "at all the sites visited, the local people were aware of the pending decision concerning the Three Gorges Project" and "the local leaders have been and will continue to be involved in the planning of resettlement within their own communities."[52] The "local leaders" CYJV claims are involved in the resettlement planning probably refer to Communist Party cadres (*ganbus*). Cadres are not elected representatives – they represent the interests of the central government rather than the local people. Under the repressive political regime in China, any local people critical of the Three Gorges Project could hardly be expected to voice their opinions. There are no nongovernmental organizations or grassroots movements, such as the committees of "oustees" threatened by India's planned Narmada dams, who would be in a position to question official decisions.

The author's personal impressions from an eight-day visit to the reservoir region in 1987 do not confirm CYJV's assertion that the local population has been widely consulted. Those people this author met who were aware of the proposed dam certainly did not share CYJV's view that resettlement "offers an opportunity for planned regional development that can add to the benefits of the Three Gorges Project."[53]

Secrecy, Irreversibility, and Bias

The inadequacy of the CYJV feasibility study demonstrates the need for public scrutiny and debate about not only the Three Gorges Project, but all development projects. The secrecy that has surrounded both China's drafting of the Three Gorges proposal and Canada's feasibility study has denied these efforts much valuable (and free) input from people beyond the confines of the institutions entrusted with the studies.[54]

Secrecy results in important factors being overlooked in the various planning and decision-making stages, and also prevents correction of omissions or distortions before the project becomes "irreversible." In fact, the presentation of development plans as "irreversible" is often a deliberate strategy to avert more rigorous discussions of projects which, in cases such as the Three Gorges Project, would likely be rejected if the true costs and benefits were brought to light.* The CYJV feasibility study's language indicates that the

---

*This is the case for Brazil's infamous Balbina Dam, which was built on the Uatumã River in 1987. Before construction began, the government continually referred to the project as "irreversible," yet the costs of the dam so greatly exceeded its benefits that money could have been saved by abandoning the scheme even after the river had been blocked and the reservoir was partially full.

authors accept as a foregone conclusion that the Three Gorges Dam will be built along with other grandiose megaprojects affecting the Yangtze River, such as China's interbasin water transfer schemes. Specifically referring to these developments, CYJV states flatly that "the Yangtze watershed *will be* [author's emphasis] subjected to many human interventions over the next several decades."[55]

The CYJV study team appears not to have fulfilled its role, at least theoretically included in its mandate, which was to provide input to the decision on whether or not to build the Three Gorges Dam.

Chapter Four

Three Gorges Reservoir: Environmental Impacts

by David L. Wegner, M.Sc.

Background

A reservoir is an impounded body of water created when a river or stream is dammed and water is allowed to store. This impoundment of water has an immediate impact on the physical and biological systems within the reservoir which needs to be understood before the full range of environmental impacts can be properly evaluated.

CYJV recognizes that the impoundment of the Yangtze River to create a reservoir hundreds of kilometres long would cause environmental impacts affecting physical, ecological and social change within the river basin and tributaries. It states that water velocity in the reservoir would be reduced to 20 percent of the river's natural velocity at certain times of the year, and that the reservoir would disrupt aquatic and terrestrial (land-based) ecosystems. However, CYJV fails to assess the range of impacts that could occur, and instead depicts a reservoir that would always operate within very narrow, average conditions.

From River to Reservoir
Physical Changes

Of particular concern is CYJV's failure to identify and evaluate the existing range of physical conditions in the reservoir. Based on the experience with reservoir impoundments elsewhere, it is reasonable to expect that the Three Gorges reservoir would initiate the following physical changes in the river system:

- The free flowing river would be transformed into a slower moving, or still water system.
- Migration of fish either upstream or downstream would be blocked by the dam.
- As water velocity is reduced, fine particles of sand and clay, known as sediment, would settle to the bottom of the reservoir.
- Initially, as the water level in the reservoir rose, land would be inundated and nutrients (and pollutants) would be drawn out from the flooded soil and decomposing vegetation.
- The flooded vegetation would provide new habitat for young fish.
- Evaporation of water would increase.
- Temperatures in the upstream end, and in the upper portions, of the reservoir would increase.
- Due to the combined influence of temperature, sediment and other chemical constituents, the water in the reservoir would become stratified. That is, the water body would not be evenly mixed, and this would have a wide range of results. For example, fish might be unable to use some areas because of lack of oxygen or temperature limits.
- Water clarity in the reservoir might increase as the water slowed down and stratification occurred.
- As the water seeped into the dry land, and as the reservoir level fluctuated, the reservoir shoreline would begin to erode.
- Stratification could cause some chemicals attached to bottom sediments to be released into the reservoir, leading to water quality problems and potential impacts on the aquatic food base.

Biological Changes

A biologically productive reservoir depends primarily on

the presence of a thriving population of various aquatic organisms such as plankton, zooplankton, and macro invertebrates, upon which fish populations depend.

As the reservoir fills, the flooding of land and vegetation would release nutrients and vegetative debris into the water. The increase in nutrients combined with the increased light penetration in the reservoir would cause the plankton to multiply rapidly. As the plankton food base responded to the new conditions, zooplankton and macro invertebrates would respond in much the same way, resulting in an overall short-term increase in the productivity of the reservoir ecosystem. But once the nutrients from the flooded soils were depleted, the plankton population might either decrease or increase, depending on the inflow of nutrients to the reservoir.

Fish presently in the Yangtze River have developed in a riverine system. If the river were to be transformed into a reservoir, the native fish species would, in all likelihood, be unable to maintain themselves. They would attempt to move upstream to more favourable waters or to seek other refuges. The new habitat created by the impoundment would tend to attract new fish species from upstream areas that are adapted to a reservoir-type environment. As well, exotic or commercially advantageous fish species would be artificially planted in the reservoir. The new fish would tend to dominate the reservoir and take over the ecological niches formerly used by the native species, causing a boom in numbers and biomass (number and total weight of fish) as they quickly exploited the available resources.

Once the new fish had exploited the reservoir resources there would be a decline in fish population. In the long term, whether or not fish populations could be sustained in their new environment would be determined primarily by the rate of flow through the reservoir and the amount of fluctuation in flows per year.

This phenomenon is widespread and has been well documented, particularly for reservoirs in the U.S. and the Soviet Union. For example, native fish species in the Colorado River (U.S.) are now either threatened or endangered due to the extensive development of dams and reservoirs along the river and its tributaries.

River Basin Impacts

The Yangtze River, like all rivers, is an ecological system: what happens upstream has an impact on what happens downstream; some impacts are immediate while others are more gradual. Many of the environmental problems associated with major U.S. rivers, such as the Mississippi, the Missouri, the Colorado, and the Columbia, can be largely attributed to a lack of systematic understanding of river basin impacts.

The overall basin relationships are what would ultimately define the inflowing water volume, water quality and biological responses. CYJV presents little information to provide an understanding of the river's potential changes as a system; for example:

- Historical or background information on how other reservoirs in China, or around the world, have developed as a result of reservoir impoundment.
- Specific data on nutrients, productivity, and water quality, from other reservoir developments, such as those on the Yellow River.
- Seasonal and annual water flows and volumes to be expected throughout the system. It is the ranges not the averages which define the aquatic system dynamics.

Cumulative Impacts

A study of cumulative impacts would attempt to relate the

impacts of all of the mainstem developments to the ecological relationships along the river. An assessment of cumulative impacts is necessary to identify additive or synergistic effects of planned and existing developments downstream and upstream of the TGP.

CYJV failed to conduct a cumulative impact assessment of potential impacts both downstream and upstream, and the effects of upstream water development on the planned reservoir. On a river and project of this magnitude, potential environmental impacts and concerns cannot be thoroughly assessed by studying only one component of the river system.

Environmental Relationships in the Three Gorges Reservoir

CYJV's assumptions in its ecological and environmental assessments are based on either no data or data that is outdated, non-verifiable, and representative of only a narrow band of conditions. With regard to the Chinese assessment of upstream environmental hydrology, CYJV simply notes:

> Reservoir tributaries and reservoir characteristics were not addressed as they are not significant with regard to engineering studies for the Three Gorges Project.[1]

In general terms, CYJV discusses the environmental responses to be "expected" as a result of the dam and reservoir but fails to define adequately a range of new conditions that could occur.

Reservoir Productivity

CYJV's conclusion that the change in aquatic environment would provide an increase in overall productivity of the aquatic environment is unsupported by world experience with large-scale reservoirs, and questionable for the following reasons:

- **Outdated data base**

    Although CYJV recognizes that populations of aquatic organisms such as plankton and invertebrates are important parameters for relating predicted water quality changes to overall reservoir productivity, the assessment of existing populations in the river was conducted roughly fifty years ago by the Wuhan Aquatic Institute. In reference to this, CYJV states:

    > The Chinese believe that, since river flow characteristics have not changed, the results obtained in the early 1950s are still representative of the present day situation.[2]

- **Inadequate analysis of river flow variability**

    CYJV fails to provide adequate information on a range of variables which form the basis for conducting a thorough and comprehensive analysis of the effects of flow variability* on reservoir productivity. In particular, hydrological variability typically would affect retention time for water in the reservoir, and hence, reservoir productivity. Rather, CYJV bases its discussion of the reservoir on "average" proposed river operations without considering the wide variability and range of hydrological conditions in the Yangtze River system. The use of "average" flows to describe the system is misleading in many respects because "average" conditions occur only in a textbook. In reality, a

---

*As an example of the high flow variability in reservoirs, the Glen Canyon Dam's reservoir, Lake Powell, has experienced inflows from 8 billion cubic metres to over 26 billion cubic metres over the past five years.

reservoir ecosystem is dynamic and responds more to the variability of the system than to "average" conditions. Also, CYJV fails to consider "regulated" flow conditions due to future developments upstream which would ultimately affect flows into the reservoir.

- **Inadequate analysis of reservoir operation**

CYJV provides limited information on how the dam would actually be operated for flood control, hydropower genera-tion, and navigation. For example, it states that the reservoir would be maintained at an elevation of 140 metres for as long as possible during the flood season but fails to explain specifically how the large annual flows would be routed through the reservoir, and what the flow manage-ment priorities would be for the entire river basin.

CYJV does state that the reservoir would only be capable of storing nine days of the river's average annual flow and that there would only be a short retention period for water in the reservoir before discharging downstream. As a flow-through system, the reservoir could have large impacts on the day-to-day reservoir dynamics and productivity, but CYJV fails to evaluate the significance of this on overall productivity.

More specifically, CYJV's discussion of reservoir produc-tivity lacks adequate information on: the seasonal and annual ranges of water quality upstream of the dam site; the contribution of nutrients and pollutants from the Yangtze's upstream tributaries; and the seasonal and annual variabil-ity in regional temperatures, water supply, and nutrient loading. CYJV also fails to analyze how the sediment would move through the reservoir, which would have a significant impact on how productivity in the reservoir would develop.

Clearly, there is little understanding of how the reservoir would react to the range of new conditions. But based on experience with other reservoirs, and CYJV's description of the reservoir as a flow-through system with a short retention time, productivity in the reservoir would tend to be limited – contrary to CYJV's prediction.

Fisheries

Currently, there are 172 species of fish in the reservoir region, 25 of which are caught commercially. These fish, like the common carp which comprises up to 18 percent of the total commercial catch in the region, are adapted to the rapidly flowing river, and spawn from April to June as water levels in the Yangtze rise.

CYJV reports that the reservoir would flood existing aquaculture facilities, which currently produce twenty times more than the annual natural fisheries. The irrigation ponds and rice fields used for raising fish would all be flooded.

To estimate aquacultural losses, two of the 14 counties that would be affected by the reservoir impoundment were surveyed. CYJV reports that the two main fry production facilities in those two counties, Wanxian and Fengjie, would be flooded. On the basis of this survey, CYJV then estimated that the reservoir "could eliminate over 3,888 tonnes of fish production at an estimated value of 15.5 x 10⁶ yuan per year [$521,750]."[3] Incredibly, CYJV predicts elsewhere in its report that aquacultural harvests in the reservoir area could increase from 54 to 109 percent.

For the effects of the reservoir on the natural fisheries, CYJV states without substantiation that the natural-catch fishery would increase by 59 percent. This statement is misleading for the following reasons:

- The expected heavy build-up of sediment in the reservoir would likely have a negative impact on the spawning ability of the fish such that the reservoir would have to

be stocked – CYJV acknowledges this.
- The long-term decline in reservoir fisheries is well documented for many river systems around the world. If the Three Gorges reservoir were stocked, as is proposed by CYJV, native species would decline in numbers as the exotic species dominate. This would be followed by a slow reduction in the genetic quality of the natural or native fish populations, resulting in an eventual loss of fish populations, and of diversity and fish health.
- CYJV fails to adequately define biological productivity and natural fish populations, potential modifications to habitat, and modifications to water quality due to upstream developments. All this is necessary to predict which fish species would increase or decrease, and what level of stocking would be required.

Overall, it appears highly unlikely that the natural fisheries would expand, and far more likely, based on the information presented by CYJV, that they would suffer a serious decline.

Other Reservoir Basin Impacts Not Considered
- CYJV fails to assess how reservoir shoreline erosion would affect biological conditions (for example, fish spawning and terrestrial habitat) and human use of the reservoir, and conversely, how the physical features of the river banks and shoreline would be affected by reservoir operations.
- CYJV assumes that sedimentation at the mouths of tributaries upstream of the dam would restrict the ability of fish to migrate back to their spawning grounds. No definitive discussion is provided on the upstream sources of sediment, the impact of sedimentation on fish reproduction, or measures that should be taken to reduce the river's sediment load.

- CYJV fails to recommend a program to maintain slope stability around the reservoir in order to minimize beach erosion and stabilize the reservoir basin. Substantial slope failure could reduce the reservoir area and have an adverse impact on productive aquatic areas.
- CYJV fails to evaluate potential water quality problems, pollution, and heavy metal accumulation in the reservoir as a result of existing and future upstream land use, hydrological changes and industrial activity.

Impacts on Terrestrial Communities

A number of rare wildlife species (e.g., the clouded leopard, macaque, and tufted deer) are still sighted occasionally in the reservoir area, but most land has been cleared and intensely cultivated for many years. Except for the hilly areas along the Yangtze River and a few remaining tracts of forested lands, the only significant tracts of natural habitat are found along the tributary valleys, such as the Daning River Valley. CYJV acknowledges that the reservoir would change the ecological and social conditions in the tributary valleys but concludes:

> Information on these tributaries is insufficient to assess the significance of flooding and resettlements on wildlife habitats.[4]

The limited discussion of the remaining natural habitat and wildlife in the reservoir area neglects the following:

- effects on the various species in the reservoir area;
- the number of acres of riparian vegetation which would be lost to the reservoir;
- potential erosion of the riparian corridor along the 2500-kilometre reservoir shoreline; and

• links between terrestrial communities and local people.

Since certain critical areas would have to be protected during and after construction, CYJV should have provided a map showing sensitive, critical, and developed areas.

Conclusion

Lacking key information on environmental hydrology, cumulative impacts, biological, physical, and chemical responses, and human use patterns, it is impossible to truly assess the impacts of the Three Gorges reservoir. There is enough doubt in CYJV's data to warrant a much more extensive assessment of the potential impacts of the Three Gorges development.

Chapter Five

Potential Methyl Mercury Contamination in the Three Gorges Reservoir

by Alan Penn, M.Sc.

Background on Methyl Mercury Contamination

The CYJV study indicates that inorganic mercury is present in the sediment, soil, and vegetation that would be subject to flooding by the Three Gorges Project. Canadian experience with the impoundment of rivers to create reservoirs for hydroelectric dams demonstrates that methyl mercury, a central nervous system toxin, is formed through bacterial synthesis in flooded soils and vegetation. The methyl mercury so produced is accumulated by fish, and is thus a potential health hazard for consumers of fish.

The process of methyl mercury production occurs naturally in lakes and rivers but certain new reservoirs in northern Manitoba and northwestern Quebec in Canada have resulted in a four to six-fold increase in methyl mercury concentrations in fish. In particular, methyl mercury contamination has become a significant issue at the La Grande hydroelectric complex (Phase I of the James Bay Project) which was built by Hydro-Québec in northwestern Quebec between 1970 and 1984, and will continue to play a role in the Quebec government's plans to proceed with additional hydroelectric development in the same region.

Prior to hydroelectric development in these regions of Canada, methyl mercury contamination was a concern because of regionally elevated levels of mercury found in fish. In the 1970s, methyl mercury concentrations as high as 100 milligrams per kilogram (mg/kg) were found in the hair of some Cree Indian fishermen. The World Health Organization sets the tolerance limit for human exposure to methyl mercury at 6 mg/kg; therefore, the high levels found in the Crees have prompted questions about long-term toxicity and the possible effects of fetal exposure.

Since the construction of Phase I of the James Bay Project, limits have been placed on fish consumption in order to control exposure to methyl mercury. Recent surveys of adult Crees over 40 years old reveal that approximately 5 percent of this adult population have in excess of 25 mg/kg in their hair.

In the case of the Three Gorges Project, CYJV has identified methyl mercury contamination as a potentially major water quality impact and proposes a monitoring program for soil, vegetation, and fish – including more scientific studies, post-impoundment monitoring of fish, and restricting consumption should mercury levels in fish become a health risk.

CYJV suggests several conditions which would act to reduce the extent of methyl mercury contamination: high annual flows through the reservoir; low organic matter in soils and absence of peat in the reservoir region; removal of vegetation before flooding; and the water's resistance to acidification. Factors which would act to increase the likelihood of contamination are cited as dissolved oxygen depletion in the reservoir, mercury present in soils and vegetation, and the transport of mercury with suspended sediment.

Key Criticisms of CYJV's Discussion of Methyl Mercury

• In the past two decades, methyl mercury contamination associated with North American water resource development projects has stimulated research on the factors which influence the rate of methyl mercury production and breakdown over flooded terrain, and its subsequent accumulation by fish. CYJV does little to assess the extent to which

Canadian or United States research on methyl mercury in the aquatic environment could be used in the evaluation of the Three Gorges Project. Without further documentation on the nature of available research, the CYJV text on this subject must be considered inadequate, and potentially misleading.

- In any event, applying experience from water resource development in Manitoba and Quebec to the very different geological setting of the Three Gorges Project has its own limitations. Even in the absence of flooding to create a reservoir, current understanding of geochemical factors which influence the bioavailability of inorganic mercury and production of methyl mercury is quite limited. These technical limitations which make impact assessment in this area rather uncertain, were not brought out by CYJV in its study.

- The rate of accumulation of methyl mercury by fish ultimately depends on the feeding ecology (i.e., feeding patterns and habitat) of the fish species present and the nature of the food chain. There is little in the CYJV study to indicate how these biological factors would influence the nature of anticipated methyl mercury contamination in the case of the Three Gorges Project.

- CYJV did not conduct a risk assessment for plausible scenarios of methyl mercury contamination in consumers of fish from the Three Gorges reservoir. The current health status of the population and the role of fish in the diet (actual or potential) are not addressed. Also, the role of fish as a subsistence food within the local economy is unclear. In these circumstances, it is especially difficult to assess the extent to which methyl mercury contamination should be considered a significant public health issue. There appears to be no fundamental reason that such a risk assessment could not have been undertaken.

- CYJV's treatment of methyl mercury as an environmental issue in the feasibility study, not unexpectedly, reflects Hydro-Québec's approach to, and experience with, Phase I of the James Bay Project. Hydro-Québec's guiding principle* seems to be that ecological impacts, by their very nature, are difficult to predict. Therefore, the argument runs, rather than rely on prior assessments, ecological and social consequences should be monitored as they unfold and responded to by the proponents, where necessary, with appropriate remedial measures. This leaves a great deal of discretion and control in the hands of proponents and places local people, who are directly affected by ecological impacts, in a disadvantaged position.

- In this setting, this also means that it is difficult to deal directly with ecological concerns, such as methyl mercury contamination, in the initial project design. A more fundamental concern is that the overall approach to hydroelectric development reflects an underlying lack of conviction about the relevance of initial environmental impact assessments and about the value of acquiring practical experience through those assessments before decisions are made.

---

*Hydro-Québec's approach is set out in Section 8 of the James Bay and Northern Quebec Agreement, the aboriginal land claims settlement associated with the James Bay Project.

Chapter Six

Downstream Environmental Impacts

by Joseph S. Larson, Ph.D.

The impacts which may occur downstream do
not affect the overall environmental feasibility
and may indeed enhance the environment.[1]

This chapter criticizes the adequacy of the CYJV feasibility study with respect to its assessment of downstream environmental impacts. The impact of dams on river systems has
all too often been poorly understood, which has led to costly
and irreversible changes which have had a dramatic impact
on local and regional economies, resources and people.

Downstream of the Three Gorges, there are six areas
which would be subject to impacts: the middle reach of the
Yangtze River between the existing Gezhouba Dam at Yichang
down to Wuhan; Dongting Lake; the Yangtze mainstem
between Wuhan and Datong; Poyang Lake; the estuary; and
the coastal/marine area.

- **The middle and lower reach** extends 1170 kilometres
  from Yichang to Datong. An estimated 75 million people live
  in this region of the Yangtze River basin, which encom-
  passes roughly 60,000 square kilometres. Based on its
  environmental characteristics, the middle and lower reach
  area is further subdivided as follows:
- **The middle reach** meanders 700 kilometres from Yichang
  to Wuhan, cutting through a wide and low-lying fertile
  floodplain known as the Jingbei Plain. With an average
  depth of 10 metres, the river channel is completely dyked in
  to protect the surrounding area from flooding. Roughly
  30,000 kilometres of dykes form part of an extensive flood

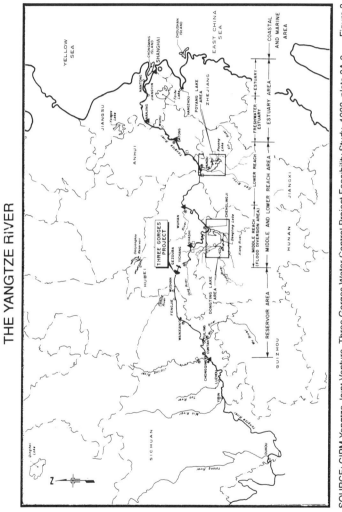

THE YANGTZE RIVER

SOURCE: CIPM Yangtze Joint Venture, Three Gorges Water Control Project Feasibility Study, 1988, plate 8A-2. Figure 3

protection system throughout this region.

- **Dongting Lake** in Hunan province acts as a natural flood storage reservoir for the Yangtze River and is strongly influenced by flows and water levels in the Yangtze River as well as other tributaries flowing into it.

- **The Yangtze mainstem**, extending 500 kilometres from Wuhan to Datong, is flanked by a large, fertile alluvial plain. Broader and shallower than the middle reach, varying from three to seven metres in depth, the river divides into several channels and islands. Wuhan is the final inland port accessible to ocean-going vessels.

- **Poyang Lake** in Jiangxi province is influenced by the seasonal high-water levels in the Yangtze River as well as in the rivers in Jiangxi province which flow into it. When water levels are high in the Yangtze, Poyang Lake is unable to drain into the Yangtze, and the lake spreads out over the surrounding plains.

Yangtze River Hydrology

To understand the impact the Three Gorges Project would have on downstream areas, the hydrological characteristics governing the river system must first be understood. The volume, timing, periodicity, and duration of water flows, and the dissolved and suspended soil sediment carried, dictate the fate of downstream fisheries, farming, water quality, wildlife and estuarine life.

The extent to which the Yangtze River influences water levels downstream of Three Gorges depends on a number of factors: flood flows from the upper reaches, timing of regional monsoon precipitation, and tributary flood patterns in the downstream lake region. During the rainy season, the two floodplain lakes, Dongting and Poyang, accept back flow from the Yangtze. But when local flooding coincides with the Yangtze's peak flood flows, water is diverted to the vast

Jinjiang Flood Diversion Area and a huge flood spreads over the land.

CYJV bases its assessment mainly on the 1985 Chinese environmental impact statement which is seriously flawed because it considered hydrology only as it related to flood control, power generation and navigation; downstream environmental hydrologic characteristics are not documented.*

CYJV reports that the occurrence of major floods cannot be predicted, pointing out that many decades may pass without serious floods and then several may occur within a few years. It notes that reservoir operations at the dam would change water levels in the middle and lower reaches of the Yangtze, which would, in turn, modify water levels in the mainstem, and in Dongting and Poyang Lakes. In the flood season, daily flood flows would be significantly reduced, and in the dry season, water levels would fluctuate widely on an hourly and daily basis. However, CYJV lacks the data required to determine the actual impact of changes in river levels.[2]

Yangtze River Sediment

The transport and deposition of the river's nutrient-rich sediment is second only to hydrology in determining the basic character of the aquatic environment immediately downstream of the dam to the estuary 1850 kilometres away. In the Chinese environmental impact statement, sediment was discussed only in terms of sediment management in the reservoir.

---

*CYJV states that its assessment is based on the Chinese environmental impact statement, as well as on data from the YVPO Preliminary Design Report and information from international sources. However, the author was unable to verify the documentation used by CYJV because prior to the release of the CYJV study, CYJV deleted various sections of Volume 8 pursuant to Section 19(1) of the Access to Information Act.

Initially, the reservoir is expected to trap 60 to 70 percent of the incoming sediment in the reservoir; then after roughly 100 years the reservoir is expected to reach a state of equilibrium where sediment is neither accumulating nor being flushed out of the reservoir. However, the lack of a comprehensive analysis of sediment processes means that the impact of a reduced sediment load remains poorly understood.

Without documentation of the relationship between Yangtze hydrology and sediment processes, and downstream ecosystems, CYJV's impact assessment is inadequate, particularly with regard to the following impacts:

• **Channel degradation**

Once the river loses its sediment to the reservoir, it would gain more power to erode its channel and banks downstream. CYJV expects the increased erosion downstream could cause the river to shift its course, threatening all flood protection dykes and low-lying river banks, thereby increasing the risk of a disastrous flood along the middle and lower reaches.

An additional problem, which CYJV suggests could increase channel degradation, would be the mining of sand and gravel immediately downstream of the Gezhouba Dam. The river would lose a source of sediment which it normally scours and carries away during flood flows, thereby causing more of an erosion problem further downstream.

Yet despite the potential dangers, the available hydrologic data are not sufficient to make a confident prediction with respect to this threat, and it is unclear whether the cost of mitigating channel degradation, either in the short term or long term, has been included in the cost-benefit calculations.

• **Hazardous pollutants**

Operation of the reservoir could induce chemical changes

near the reservoir bottom which would release hazardous pollutants such as heavy metals, chemical pesticides, and fertilizers. Suspended sediment particles in the water could distribute the hazardous pollutants downstream. CYJV notes that the adsorptive capacity of the sediment particles must be known in order to determine whether pollutants would accumulate in the reservoir or be dispersed downstream, and to predict the impact this would have on downstream water quality and fisheries. Despite the importance of this issue, CYJV merely states that the Chinese government has evaluated the adsorptive capacity of the three main types of sediment which would be found in the reservoir. It does not discuss the results or their impact on downstream water quality and fisheries.

• **Gas bubble disease**

The spillways which would be installed at the Three Gorges Dam are known to cause gas supersaturation (water supersaturated with air), which produces a fatal condition in fish and other gilled animals, which is known as gas bubble disease and is similar to diver's bends. CYJV reports that the Chinese government is aware of the potential danger, but has dismissed it because their investigations at the Gezhouba Dam indicate little problem among downstream fish fry. The Chinese government did not provide CYJV with the details of these studies nor their methodology for verification; and, furthermore, the proposed spillways are a different design from the Gezhouba spillways.

In Volume 8 on environment, CYJV states that it is unwilling to dismiss gas bubble disease as a major impact downstream; and yet, in the study summary (Volume 1), it suggests that losses in natural fisheries and aquaculture could be compensated for monetarily or by "enhancement measures" (presumably this means stocking). No financial, biological, or engineering data is provided to support this.

Rare and Endangered Aquatic Life

- **The Chinese sturgeon**, cut off from its traditional spawning grounds upstream of the Three Gorges, now spawns in the rapidly flowing water 10 kilometres downstream of Gezhouba. CYJV acknowledges that sand and gravel mining planned for this area could adversely affect the sturgeon, but that the necessary studies to examine project effects on this species have not been done. It suggests that sturgeon hatcheries could be useful to replace lost spawning grounds and to minimize the impact of gas bubble disease, but it is not clear whether this cost has been included in the cost-benefit calculations.

- **The Chinese river dolphin**, the rarest freshwater dolphin in the world, is found only in the Yangtze's middle and lower reaches and at the confluence of Dongting Lake. According to CYJV, there are only 200 to 300 of these dolphins remaining and the expected increase in erosion downstream of the dam could adversely affect them and the semi-natural dolphin reserves established by the Chinese authorities along the Yangtze.* The Chinese environmental impact statement did not address the impact of the Three Gorges Project on the river dolphins. CYJV recommends more studies and that all means available to conserve and protect this species be considered appropriate.

- **The finless porpoise** is commonly found along harbours and bays in the coastal area and up the Yangtze into

---

*Editors' Note: The World Wide Fund for Nature (WWF) International designates the Yangtze River dolphin, or baji, as one of the most endangered species on earth. In 1992, the International Association for Aquatic Animals' Medicine named the baji – with a population of between 150 and 200 animals – the most endangered dolphin in the world.

Dongting Lake. CYJV predicts that the Three Gorges Project would have little impact on the porpoises – a judgment apparently based on the assumption that the porpoises have sufficiently adapted to a wide range of environments, and would therefore adjust to the environmental change caused by the dam.

• **The Chinese alligator** is found only in the lower reaches of the Yangtze. Only 300 to 500 of these alligators remain, inhabiting irrigation and storage ponds, rice fields and shallow depressions in low-lying plains. CYJV optimistically concludes that the impacts of the Three Gorges Project on this species would be minimal, and that the real threat to this species is ongoing harvesting at a rate that will soon cause extinction.

Dongting and Poyang Lakes

These shallow low-lying lakes are the largest freshwater lakes in China, known to support significant and productive fisheries, as well as aquaculture facilities. Despite this, the Chinese government failed to provide any substantive information concerning their aquatic environments and ecology.

• **Dongting Lake** is China's second-largest freshwater lake and can fluctuate from 600 square kilometres in the dry season to 2800 square kilometres in the flood season. A nature reserve, adjacent to Dongting Lake, provides excellent habitat for birds. Dongting Lake acts as a natural flood retention area and a depository for 20 percent of the Yangtze's sediment load carried down through the Three Gorges from the upper reaches. Over the past thirty years, the lake's surface area has been reduced by half, losing much of its storage capacity to land reclamation and sedimentation. CYJV reports that little is known about the dynamics of sediment transport and deposition through the lake system.

• **Poyang Lake** is the largest freshwater lake in China and fluctuates from under 1000 square kilometres in the dry season to 4000 square kilometres in the flood season. Water, laden with nutrient-rich sediment from the Yangtze River and five other rivers, provides large areas of rich shallow water (water 10 to 20 centimetres deep appears to be critical for cranes) and mud flats* with abundant aquatic vegetation. The Poyang Lake Nature Reserve (22,400 hectares), adjacent to the lake, provides excellent bird habitat. In 1986, the reserve wintered over 95 percent of the world's population of the rare and endangered Siberian crane, plus several other unique or rare crane species.**

CYJV notes that the Three Gorges Dam would change the water levels and the circulation and deposition of sediment at Poyang Lake, but the study's lack of quantitative information makes it very difficult to assess the effect of the dam operation on crane habitat. Certainly, water level changes in the growing season would affect the vegetation of the habitat. In Volume 8, CYJV was unable to estimate what the effects would be and concluded that a significant potential impact could occur; in Volume 1, CYJV downplays this concern.

Downstream Agriculture, Fishing, and Aquaculture

Editor's Note: The author's observations in this section are

---

*Mud flats are those areas along a watercourse or surrounding a body of water which become exposed during periods of low flow. They provide a rich food source and critical habitat for many aquatic birds.
**CYJV estimates the total number of bird species in the reserve at 90 but does not indicate when its count was taken. As of 1986, WWF-Hong Kong put the figure at 236 species. Examples of some of the larger species which depend on the lake habitat are the white-naped crane, hooded crane, common crane, white stork, white spoonbill, swan goose, and grey heron.

derived in part from a 1986 visit to Poyang Lake and travel by boat on the Yangtze between Juijang, near the confluence of the lake and river, and north to Wuhan. The author also draws on data from the 1990 Discussion Draft Management Plan Jiangxi Poyang Lake National Nature Reserve prepared by D.S. Melville, Director of Conservation, World Wide Fund for Nature-Hong Kong, Jiangxi Provincial Government.

According to CYJV, natural fish harvests have declined in the Yangtze River by roughly 50 percent since the 1950s. And since the Gezhouba Dam was built in 1981, egg production in several downstream reaches of the Yangtze has dropped by at least 50 percent. If the Three Gorges Project is built, CYJV states that changes to Yangtze flows during the months of April, May and June, when most fish are spawning, could adversely affect fisheries downstream of the Three Gorges. It further states that this issue cannot be resolved, since information is inadequate. In any case, it suggests that other impacts such as overfishing, pollution and the blocking of spawning migrations into dammed Yangtze tributaries would mask any changes caused by the Three Gorges Project.

Although CYJV acknowledges that the Three Gorges Project would affect "the human environment of some 2500 kilometres of the Yangtze River valley," it makes only passing reference to the 75 million people living along the middle and lower reaches and the effect on traditional local agriculture, fishing and aquaculture downstream of the dam. For the sake of national hydropower development, both the Chinese government and CYJV have effectively ignored the subsistence economies along the Yangtze and around the lakes. These economies could be seriously disrupted and there are few, if any, funds available in China to compensate people for their loss of livelihood or means of subsistence.

Between Poyang Lake and the city of Wuhan, the Yangtze

River and its banks and dykes are used extensively by local fishermen and by farmers grazing livestock. Still common to the area is the age-old technique of cormorant fishing.* Villagers use the Poyang Lake Reserve, even though it is flooded to a depth of several metres for several months of the year, for a wide variety of economically important activities. For example, they cut grass for fertilizer, grow subsistence crops, plant trees, fish, plant and cut reeds for paper making, and collect medicinal herbs to sell.** Around the reserve, villagers tend about 2500 water buffalo in large herds of up to 400 animals. During the dry months these water buffalo graze on parts of the reserve. An adequate assessment of the Three Gorges Project would not ignore local economies in the downstream impact assessment.

Yangtze Estuary, Wetland, and Coastal/Marine Area

> It is not possible to totally dismiss the possibility of significant impacts occurring within the estuary on the basis of information received to date.[3]

From Datong, the Yangtze River branches out into a sprawling delta (or estuary) which stretches 655 kilometres out to sea, and forms one of the largest continental shelves in

---

*The fisherman fits the cormorant, a fish-eating bird, with a ring around the throat to prevent the bird from swallowing the fish whole; the bird is then tethered to a long line and the fisherman allows the bird to dive and capture fish; when the bird returns to the boat it is rewarded with small pieces of fish.

**Grass-cutting rights are allocated among local families who live for several months in temporary shelters on the meadows and cut grass for green fertilizer on vegetable fields. Of 39 medicinal herbs harvested in the Poyang Lake Reserve, 18 species have commercial value.

the world. Over half of the Yangtze's annual sediment load is deposited in the estuary and the remainder is carried by the ocean currents out to the coastline and continental shelf. Depending on tidal influence and the seasonal flows of the Yangtze, the river and estuary waters can be fresh, brackish or salt water.

The Yangtze River estuary is rich in biological production due to the ongoing deposition of sediment and nutrients from the Yangtze, and the presence of aquatic species with an affinity for brackish water. The estuary provides a rich food source for birds and fish and is also the spawning and nursery grounds for most commercial fish. CYJV reports that anchovies, herring, eel, crab and shrimp constitute over 50 percent of the total estuary harvest. But they also report that, according to the Shanghai Environmental Protection Bureau, fish stocks in the estuary are declining due to severe pollution, overfishing, and disruption of fish migration into the dammed tributaries.

The Three Gorges Project would alter existing flow patterns and reduce sediment deposition in the estuary. In Volume 8, CYJV suggests that sediment processes may be disrupted for roughly two hundred years after the dam is in operation, but in Volume 1 concludes that the impact of the Three Gorges Project on the sediment processes at the estuary would be relatively small. An environmental impact statement on downstream impacts should have addressed the relationship between river discharge and marine dynamics, but CYJV merely states that it found no information on this.

While several Chinese studies are currently examining the estuary and its potential uses for port and navigation development, fisheries development, land reclamation and water resource management, no study of the specific effects of the Three Gorges Project has been completed. CYJV predicts little impact on aquatic life in the estuary, but does not provide any supporting evidence.

The Chinese environmental impact statement did not consider impacts on wetlands adjacent to the river's middle or lower reaches or at the estuary. CYJV states that since there is no information on the wetland environment of the estuary, its assessment was based on field trips. However, D.S. Melville of WWF-Hong Kong, reports that the East Shanghai University has studied shorebirds, intertidal flats,* marsh vegetation and aquatic life in the region. According to Melville, the mouth of the Yangtze River and nearby Hangzhou Bay are important wintering grounds for migrant shorebirds, such as the Great Knot Sandpiper, which probably flies direct to Shanghai from northwest Australia. Any ecological changes in the Yangtze estuary could have a devastating impact on the sandpipers.

Salt Water Intrusion

From December to April, when the flows from the Yangtze into the East China Sea are low (less than 20,000 cubic metres per second) and ocean currents are strong, salt water can intrude into the estuary and tributaries, rendering the water unfit for most purposes such as drinking and irrigation. In recent years, salt water intrusion has been occurring more frequently, with disastrous consequences for agricultural, municipal, and industrial water intakes from the Yangtze for the city of Shanghai and along the coast.**

---

*Intertidal flats – mud flats exposed daily due to tidal fluctuations in estuaries – provide critical feeding areas for shorebirds during low tide and for fish during high tide.

**When the salt content of water becomes too high as a result of salt water mixing with surface freshwater and seeping into the ground-water table, irrigation has to be interrupted for a few days or even several months. In 1978 – 79, 1,333 hectares of rice crops were destroyed due to the very high salt content of the irrigation water supply. Due to the salinized water supply, Shanghai experienced an economic loss of $4,963,000.

The flow of the Yangtze River influences the pattern and extent of sea water intrusion into the estuary, and therefore there is considerable concern that the predicted reduced flows in the Yangtze would increase the risk of salt water intrusion. The Chinese environmental impact statement dismisses the potential impacts at the estuary due to salt water intrusion because it assumes the dam would have a minor impact on average hydrologic conditions in the estuary, and because of the long distance between the dam site and the estuary. Although CYJV does not expect that salt water intrusion would be a problem, it admits that more information is needed.

The coastal/marine area includes the East China Sea, the Yellow Sea, and the Bo Hui Sea, and is influenced by oceanic currents, climatic variations, weather patterns (especially typhoons), and water-sediment discharges especially from the Yangtze River. Reduced flows from the Yangtze, as a result of the Three Gorges Project, could affect pollution dispersion and dilution in this area and potentially exacerbate conditions in the coastal area where hundreds of kilometres of the coastal shoreline are already severely polluted.

Although CYJV states that more information is required to establish a relationship between river discharges and the characteristics of the estuary and open sea, it says that, in general, the project would not have any significant impact in this area.

Key Criticisms

- Because CYJV was not provided with sufficient data and/or failed to acquire the necessary data, the CYJV study is fatally flawed and is not an adequate assessment of downstream impacts.
- CYJV's Volume 8, Environment, and Volume 8A to G, Appendices, demonstrate that, with respect to downstream

impacts, the feasibility of the project is poorly understood. Therefore, major adverse and costly problems cannot be ruled out, and, in many instances, there are no grounds for assuming that technological remedies exist.

- Volume 8 of the CYJV study and its appendices provide strong evidence of these inadequacies; but Volume 1, which is intended to stand alone as a summary document, is inconsistent with Volume 8 and is misleading with respect to downstream impacts.

- CYJV's credibility is in question due to its conclusion in Summary Volume 1 that engineering solutions and/or money can somehow make up for the lack of understanding of critical downstream environmental impacts.

- CYJV has identified serious omissions in the information required to evaluate and quantify environmental impacts downstream of the Three Gorges Project, which render the assessment incomplete. Despite this, CYJV glosses over serious concerns raised in Volume 8 and recommends the project as "environmentally feasible." Therefore, regardless of which agency is at fault for those omissions, CYJV's conclusion is misleading and irresponsible.

Chapter Seven

Unresolved Issues: Perspectives from China

by Shiu-hung Luk, Ph.D., and Joseph Whitney, Ph.D.

The Chinese feasibility study for the Three Gorges Project, which was conducted under the aegis of the State Planning Commission,* remains a secret government document. From 1987 to 1989, while official studies were under way, numerous research papers[1] on the feasibility of the Three Gorges Project were circulated and published in Chinese journals.

During this period, the Chinese press reported that views critical of the project were under-represented in official documents and that a vast number of critics had not been heard by decision makers in the upper echelons of China's central government. Since then, several collections of essays such as those edited by Tian Fang and Lin Fatang, senior officials with the State Planning Commission, have been published in China as a forum for dissenting views.[2]

This chapter reviews the key unresolved issues that have been raised within China about the Three Gorges Project and have not been adequately addressed by the CYJV study.

Resettlement

The scale of resettlement required for the Three Gorges Project surpasses that of other major dams such as the Sanmenxia Dam on the Yellow River, the Danjiangkou Dam on the Han River, and the Wujiangdu Dam on the Wu River. China's record on resettlement is tragic: according to China's

---

*The State Planning Commission plays a central role in China's energy and economic planning, and reports directly to the State Council.

Ministry of Water Resources,* 30 to 40 percent of the 10 million people who have been relocated to make way for hydroelectric dams since the late 1950s are still impoverished and lacking adequate food and clothing.[3] Although China has recently improved its guidelines for resettlement in accordance with the World Bank's criteria for "successful resettlement," the people who would be displaced or affected by the Three Gorges have no guarantee they would be spared the hardship and suffering associated with such schemes.

Proponents of the Three Gorges Project are well aware of the potential for social upheaval and conflict, so it is disturbing that none of the writings reviewed, either Chinese or foreign, present any evidence that the people who would be affected by the scheme have been consulted about the impact resettlement would have on their lives.[4]

The Chinese resettlement plans, as reviewed by CYJV, suggest that the displaced population could migrate uphill, so that even though they would be forced to abandon their town or village, they could remain in the same county.[5] On paper this may seem reasonable, but after examining the conditions in the upland areas, it is obvious that this scheme would not be successful.

Land Availability

Fundamentally, the problem is that the best land in the area is in the valleys which would be flooded by the reservoir. This land is already 15 percent overpopulated,[6] and the remaining land is further uphill, too steep to cultivate properly, and relatively infertile. In fact, Chinese soil scientists have estimated that five times the area of less productive uphill land would be needed to replace the 26,800 hectares of prime

---

*The Ministry of Water Resources and Electric Power has been divided into two distinct ministries, the Ministry of Water Resources and the Ministry of Energy.

agricultural land which would be lost to the reservoir.[7] This amount of land is simply unavailable in this area.

In addition to the problem of finding replacement land, Chinese critics have doubts about the plans to integrate resettlement with natural resource development as a means of creating jobs for the non-agricultural sector.[8] They question not only the economic viability of some of the proposed schemes, but whether the necessary capital and resources would ever be made available.[9]

Development Plans

For example, the proposed salt mine development in the Three Gorges region would probably have difficulty competing with salt mines that are well established elsewhere in the province. Vague plans to develop tourist industries conflict with plans to expand polluting industries in the same region. And the proposal, endorsed by CYJV, for intensive orange and dairy production appears similarly unworkable.

Resettlement Costs

- According to CYJV, over one-half of the land slated for rural resettlement is situated above 800 metres elevation. Because of the cost of access and the fact that cultivation is limited to a small range of crops at higher elevations, development there would be more expensive. CYJV apparently failed to include this cost in its estimates.
- CYJV assumes the cost of relocating nearly a dozen cities and scores of towns, along with all their basic infrastructure such as roads and water supply systems, in the Three Gorges region by assuming it would cost about the same as rebuilding Tangshan, a northeastern city, with a population of 1.4 million, after it was totally devastated by an earthquake in 1976. But state planning officials, Tian and Lin, point out that this is an underestimation because Tangshan was

rebuilt on level ground compared to the rugged hilly terrain in the Three Gorges region.[10]

The Impact of Upstream Land Use Changes on Soil Erosion, Sedimentation and Flooding

Editor's Note: The rate of sedimentation in the proposed reservoir during its lifespan depends, in part, on land use and soil erosion patterns upstream. Quantifying the actual long-term rates and patterns of sedimentation for a number of decades is a complex issue which has long been a source of contention among scientists. Here the authors explain why proponents wrongly discount the impacts of sedimentation which could significantly shorten the useful life of the Three Gorges Dam.

Numerous Chinese authorities[11] report that soil erosion, and therefore the sediment load in the Yangtze River, is increasing because of population pressure and land degradation upstream of the Three Gorges. If the Three Gorges Project is built, some scientists are predicting a 15 percent increase in expected rates of sedimentation during the reservoir's lifespan, and others are predicting even higher increases.[12] The impact of resettlement alone – due to activities such as land clearing, cutting down trees for fuel, mining, and extraction of building materials – is expected to cause an annual 2.5 percent increase in the Yangtze's sediment loads.

CYJV, on the other hand, suggests that no obvious increases or decreases in the Yangtze's sediment load have been observed, and it makes the overly conservative assumption that sediment would deposit in the reservoir at a fixed rate over time. In addition, CYJV's estimate of the amount of sediment which would become trapped in the reservoir appears low for several reasons. First, it assumes that sediment which is currently building up behind dams on upstream tributaries will remain there indefinitely. But within a few decades or less, these reservoirs will become clogged with sediment, at which time the dams will have to be taken out of operation in order to flush out the accumulated sediment. From there, the sediment will eventually be flushed into the Yangtze River, causing a significant increase in the river's total sediment load.

Yangtze Valley Planning Office reports suggest building even more dams upstream as a strategy to substantially reduce sediment input to the Three Gorges reservoir.[13] Technically, this would reduce sedimentation in the reservoir for a short term but it would eventually cause sedimentation problems further upstream. Furthermore, to build more dams to control sediment would ignore upstream land degradation and soil erosion which are the cause of the high sediment load problem.*

Secondly, CYJV assumes that only 20 percent of soil eroded from land upstream of the Three Gorges ever reaches the Yangtze River. This discounts a significant amount of eroded soil, which is first transported by floodwater and deposited onto flood plains and other low-lying areas, but would eventually be flushed into the Yangtze.

Proposed Sediment Management Strategy

Based on its initial assumptions about rates and volume of sedimentation, CYJV predicts that in the early years of dam

---

*Proponents of the project also argue that there are advanced technologies available to reduce soil erosion although they tend to be too expensive. Less expensive soil erosion control measures, such as terracing, planting grass, shrubs, and trees, would be ineffective because any planted vegetation is likely to be denuded by people in desperate need of fuelwood for cooking. In any case, proponents do not include the cost of implementing soil erosion strategies, necessary for reducing sediment input to the reservoir, in the project cost estimate.

operation, 60 to 70 percent of the river's sediment would be trapped in the reservoir. The coarser sediment would deposit in the upper end of the reservoir, known as the backwater reach, gradually forming a delta which would encroach on the dam's useful storage capacity. Also, the river channel would become raised, thereby increasing upstream flood levels and obstructing navigation, particularly in the dry season. According to CYJV, the reservoir slope would become flatter and reach a state of equilibrium after about 100 years. CYJV claims that at that time, there would be no net additional deposition or erosion of sediment in the reservoir. Moreover, CYJV believes that "If the reservoir is operated as proposed, about 90% of its effective storage can be preserved indefinitely."[14] To prevent unwanted sediment buildup in the reservoir and thereby preserve long-term storage capacity in the reservoir, water levels would be lowered to the flood control level (FCL) of 140 metres during the flood season. At this time the flow is carrying most of its sediment load, so water is released rather than stored to avoid sediment deposition. After the flood flows when the water is relatively sediment-free, the reservoir would be raised to the normal pool level (NPL) of 160 metres and maintained at that level as required for power generation. In other words, the operating rule would be to store water when clear and release when turbid.

Critics suggest that this proposed operating procedure would not be effective for several reasons:

- Although some of the finer sediments would be flushed out, this operation would have no effect on coarser sediments which are expected to form a delta beginning several hundreds of kilometres upstream.
- At the reservoir backwater, 600 kilometres upstream of the dam, floodwater would continue to deposit sediment as they flow into the reservoir, quite independently of how the dam is

operated. And as more sediment accumulates, the rate of build-up increases, extending deposition further and further upstream.

Proponents, on the other hand, suggest that this problem would be limited because at the end of each dry season, water levels in the reservoir would be lowered, allowing the river to erode the sand bars and shoals which are formed during each flood season (this process is known as retrogressive scouring).[15] In practice, however, retrogressive scouring has been ineffective hundreds of kilometres upstream of dams such as the Sanmenxia Dam on the Yellow River.[16] And in the case of the Three Gorges Project, the reservoir is nearly five times the length of the Sanmenxia reservoir, which makes it even more doubtful that the river would effectively scour sediment deposits away. To make matters worse, the narrow Tongluo Gorge, located 15 kilometres downstream of Chongqing, acts as a bottleneck in the river so that any lowering of water levels at Three Gorges to flush sediment through would have a negligible impact on the problem of sedimentation near Chongqing.

Backwater Sedimentation and Increased Flooding
Sun and Fang[17] believe that the city of Chongqing would face an increased flood risk because backwater sedimentation would raise the elevation of the river channel. CYJV recognizes that significant sedimentation would occur, thereby increasing the level of flooding near Chongqing, although it appears it did not quantify the amount and cost of dredging required to reduce the flood risk.

Navigation

Editor's Note: The Yangtze River is a major east-west artery of trade and commerce, and is strategically important to the economic development of southwest China. The 660-kilometre reach of the river between Yichang and Chongqing is characterized by numerous narrow gorges, strong currents and dangerous shoals. Because of this, navigation is treacherous, and the cost of shipping through this reach is more than double the cost below Yichang. According to proponents, the Three Gorges reservoir would transform this hazardous reach into a deep, gently flowing waterway, which would allow large ocean-going vessels access to the river port of Chongqing.[18] The resultant increase in shipping would, in turn, facilitate the development of Chongqing as the largest municipality and inland port in southwestern China.

CYJV defines navigation benefits as equivalent to the reduced transportation costs of moving cargo and passengers through this reach of the Yangtze. The calculation of benefits is largely dependent on, and proportional to, the projected increases in shipping traffic.

Increased Volume of Shipping on the Yangtze

Members of the Chinese People's Political Consultative Committee (CPPCC), an influential group of "opposition" parties, are sceptical about the Ministry of Communication's projected five-fold increase (an annual goal of 50 million tonnes) in shipping as a result of the improved navigation. CYJV's estimate of 41 million tonnes is equally theoretical, since no thorough study has yet been done to determine the volume of shipping traffic that could be moved through the locks under various scenarios of vessel and tow size, proportion of passenger vessels to freight vessels, and with various traffic control procedures.

Impact of Sedimentation on Navigation

CYJV expects that backwater sedimentation would

obstruct navigation in the channel and access to river port facilities near Chongqing, particularly in the dry season. CYJV failed to investigate this, and also the feasibility and cost of dredging operations which could be required on a massive scale.

Another issue not yet raised by proponents is the impact of sediment releases from the Three Gorges Project on the Gezhouba Dam which is 40 kilometres downstream. These sediment releases could form a delta where the Yangtze River slows down to meet the Gezhouba reservoir. As would be the case upstream of the Three Gorges Project, sedimentation in the Gezhouba reservoir would not only reduce its limited storage capacity but, without continuous dredging, could also impede navigation.

Navigation Benefits Achievable
Without the Three Gorges Project

CYJV provides preliminary evidence that without the Three Gorges Project the present volume of shipping traffic (6 to 9 million tonnes) could be roughly tripled (17 to 28 million tonnes) depending on the size and mix of vessels. They also state that better traffic control procedures, extended hours for navigation, more powerful tug boats, and improved barge design could all serve to increase channel capacity beyond present limits. If these improvements were implemented, the volume of shipping traffic could equal or possibly exceed CYJV's projections for the Three Gorges Project.

The Multipurpose Conflict

Quite apart from the technical problem of managing the sediment that could impede navigation through this reach, and the wide range of complex factors influencing improved transportation on the Yangtze, neither CYJV nor the various Chinese sources have dealt adequately with one of the most important issues: the inherent conflict associated with operating a multipurpose dam. Intended to generate power, provide flood control, and improve navigation, the reservoir would have to be maintained at different levels to achieve optimum benefits for each function.

Generally speaking, for power generation and navigation, the higher the water levels in the reservoir the better. Conversely, for flood control – the primary purpose of the dam – water levels should be as low as possible prior to the flood season in preparation for storing peak floodwater. To complicate the matter further, the operation for controlling sediment requires that water be stored only in the dry season when it is relatively sediment-free, and released in the flood season when the river's sediment load is highest.

Proponents claim that shipping costs for vessels proceeding upstream against the strong current would be reduced due to the slower velocities in the reservoir. But if little water is actually being stored, in order to avoid sediment buildup, then it is not clear whether velocities through the gorges would be significantly reduced. Neither is it clear what the impact of operating to serve peak electricity demands in the dry season would be on flows downstream of the Three Gorges Dam and the Gezhouba reservoir. If, for example, the flows were too low, navigation depths would be insufficient and navigation would be impeded through this section.

According to CYJV, the 160-metre recommended scheme would eliminate all but one of the 32 existing bottlenecks through the narrow gorges reach of the river and the 12 existing winching stations.* However, CYJV downplays the fact that the reservoir would, at best, be held at this level for

---

*Winching stations are situated at narrow sections of the Yangtze River. A mechanical winch and steel cable, attached to the oncoming vessel by a tugboat, is used to pull the boat through the narrow section.

only six months of the year (November to April). For the remainder of the year, the reservoir would either be at the flood control operating level (FCL) of 140 metres or fluctuating somewhere between FCL and normal pool level – assuming floodwater are not stored. When the reservoir is held at the FCL, anywhere from five to eight bottleneck sections would still exist in the 60-kilometre reach downstream of Chongqing. Proponents have failed to consider how the bottlenecks at low water levels would affect larger ships which are expected to travel the improved waterway.

Conclusions

Despite the major investment of time and effort on the part of CYJV in preparing the feasibility study for the Three Gorges Project, there remain serious conceptual and data shortcomings with respect to resettlement, reservoir sedimentation, upstream flooding, multi-purpose operation and navigation benefits. Because a large number of potential costs have not been evaluated by proponents, and are not included in the cost-benefit analysis, it is by no means certain that the benefits of the Three Gorges Project outweigh the costs. If a rigorous cost-benefit analysis of the Three Gorges Project were to include these costs, the proposed scheme would appear far less economical than its proponents now claim.

Chapter Eight
Flood Control Analysis

by Philip B. Williams, Ph.D., P.E.

Background

For centuries the Chinese people have been building earthen dykes to prevent the Yangtze River from overtopping its banks during the flood season and inundating the adjacent floodplain. Over the past four decades, the Chinese government has mobilized the people of the Yangtze Valley to substantially upgrade the existing system of flood protection dykes and diversion works. Flood storage reservoirs and control structures have been constructed, overflow diversion areas have been established, and roughly 30,000 kilometres of dykes now line the Yangtze and its tributaries, and protect urban centres situated in flood-prone areas. The main dykes protecting the valley have been reinforced and raised to an average height of 12 metres and along the critical 180-kilometre Jinjiang reach, the dykes stand 16 metres high in places. Today the Yangtze Valley is capable of safely diverting and storing roughly half of a flood equivalent to that which occurred in 1954.

Ironically, as the dykes have been raised to increase the river's channel capacity, the risk of a flood disaster has grown dangerously high. If the Jinjiang Dyke – the main dyke in the middle reach which Yangtze Valley Planning Office engineers view as the critical component of the existing system – were to fail, a huge flood would spread across the densely populated floodplain, killing at least 100,000 people, and inundating major urban centres. Conceivably, the river might change course entirely and rush headlong into the city of Wuhan with a population of six million people.

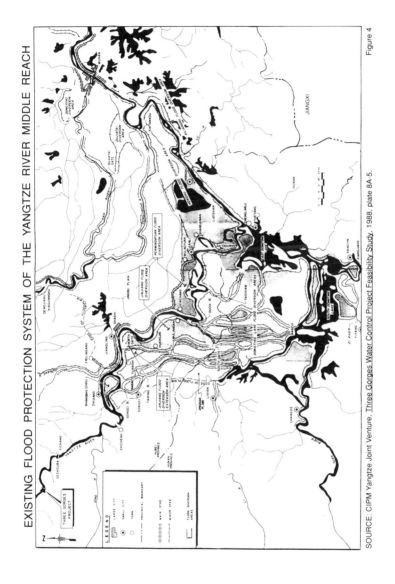

EXISTING FLOOD PROTECTION SYSTEM OF THE YANGTZE RIVER MIDDLE REACH

SOURCE: CIPM Yangtze Joint Venture, Three Gorges Water Control Project Feasibility Study, 1988, plate 8A-5.

Figure 4

- 101 -

The Existing Flood Management System

The fertile floodplain of the middle Yangtze Valley has been formed over hundreds of thousands of years by sediments deposited from the flood waters of the Yangtze and its major tributaries. Over the last two thousand years, as the population of the valley grew and farming became more intensive, a sophisticated flood management system was constructed to reduce the risk of flooding. This system has been substantially improved in the last 40 years and now consists of a network of river dykes, overflow diversion areas, control weirs, floodplain lakes and ring dykes. The guiding philosophy in the management of this system has been to attempt to absorb the huge volumes of floodwater by temporarily storing them in emergency diversion areas on the floodplain and in floodplain lakes rather than attempt to 'control' all floods.

With this system, three types of flooding can occur:

- Along the river itself, between the main river dykes, there are settlements and cultivated land. Referred to as 'beach areas' in the CYJV report, they are susceptible to flooding from frequent high river stages. Here people know they are likely to be flooded and therefore take reasonable precautions to limit flood damage. According to CYJV, flood damages can occur in beach areas if the flood exceeds the expected 5-year flood level.
- In the overflow diversion areas, covering about 800 square kilometres, about 4 million people live and farm. The use of these overflow areas to store flood waters is infrequent, and according to CYJV, occurs about once in 20 years. Although the residents of the overflow diversion areas know that they live in flood overflow areas, the level of protection varies. Many of the towns are protected by ring dykes and some areas have constructed refuges protected by dykes. Other

areas have little protection.

- Former floodplain areas protected by major river dykes have virtually eliminated flooding. Referred to as 'primary protected areas' in the CYJV report, these floodplains have been intensively developed. Of particular importance is the Jingbei Plain protected by the Jianjiang Dyke against flood levels expected only once every 250 years (or a 1:250 flood probability). The Jingbei Plain and major cities, such as Zhijiang, Shashi, and Wuhan, depend on maintaining the integrity of the 12 to 16-metre-high Jianjiang Dyke during extreme floods. The integrity can be threatened by seepage, erosion, and overtopping.

This flood management system has functioned well, significantly limiting flood damages in the worst flood of the century, the 1954 flood, which CYJV estimates to be a once in 200-year event. The flood management system was successful because of its ability to temporarily store huge volumes of water during the flood. The 1954 flood is estimated to have discharged 366 cubic kilometres of water in 60 days. At least 100 cubic kilometres were probably stored in floodplain lakes, overflow diversion areas, and in river channels.

This puts into perspective the comparatively minor role the Three Gorges Project would have on reducing flood volumes. If designed safely, and operated correctly, the dam is intended to store 31 cubic kilometres which is less then 10 percent of the 387 cubic kilometre volume of the 1000-year flood.

Review of the CYJV Flood Control Analysis

Although most of the economic benefits claimed for the Three Gorges Project are for electricity generation, the dam's advocates have maintained that flood protection is the primary need for the project. In fact, it has been frequently stated that the dam must be built to avert catastrophic flooding affecting millions of people living along the middle reaches of the Yangtze River. For a project of this magnitude – one of the world's largest civil engineering projects that could potentially affect hundreds of millions of people – the CYJV feasibility study should provide an objective and systematic analysis of the reduction in flood risk which would result if the dam were built, and evidence that the project would be the most cost-effective approach for providing additional flood protection. Moreover, such an analysis should be accurate, use the best available techniques and have a consistent methodology.

Unfortunately, the study does not do this, for the following reasons:

Purpose of the Project Is Misrepresented

Throughout the feasibility study there appears to be confusion over the main flood control goal. The study summary, which is written for potential financiers and key policy makers, states that the project's main flood control function is to protect 20,000 square kilometres of downstream floodplain inhabited by 10 million people. Design Volume 4, on the other hand, states that the project "can provide flood protection to approximately 1 million square kilometres."[1] And then Volume 7 on flood control states that the "primary flood control objective of the project and its operation must be to guarantee the integrity of the major dykes protecting the Jingbei Plain and the Jingnan Region" – an area of 6700 square kilometres.[2]

The computed flood benefits* described in the detailed technical analysis do not substantiate even the least

---

*Flood benefits are calculated as the expected reduction in economic losses due to flood damage, if the Three Gorges Project were built.

grandiose of these statements. According to CYJV, in the event of a 1000-year flood, the Three Gorges Dam would protect 4000 square kilometres from floods, which is only 30 percent of the total area that would be inundated without the project. But in the benefit analysis, even these figures are misleading. About three-quarters of the benefits claimed for the project are attributable to the reduction in the frequency, and not the elimination, of flooding of just three areas – two flood overflow diversion areas and the beach areas on the river side of the dykes. Only about 15 percent of the total economic benefits are attributable to protecting the highly populated Jingbei Plain and Jingnan Region, but the flood control storage and operation at Three Gorges is designed specifically for this 15 percent.

Only a detailed reading of the study reveals that the goals and operational design of the project are to upgrade protection for the Jingbei and Jingnan regions from the current 250-year flood level to 1000-year flood level. This means that the flood control operational design is not based on an objective cost-benefit analysis, but on a subjective judgement that these areas require levels of protection higher than the 250-year flood.

Unrealistic Operational Criteria

CYJV has recommended an operational plan for the dam – CYJV's Flood Control Operation – to release or store flood-water in the reservoir depending on water levels in the river downstream:

- During flood season, reservoir levels would be lowered from the normal pool level (NPL) of 160 metres to the flood control level (FCL) of 140 metres above sea level. Flood storage would be provided above 140 metres and moderate floods up to the 50-year flood level could be stored without the reservoir exceeding the NPL. For greater floods, the reservoir would rise higher depending on the magnitude of the inflows and the water level in the main channel downstream.

• In the event of a 100-year flood, water would be stored in the reservoir so that the water levels in the middle reach (at Shashi) would not exceed 44.5 metres. As such, diversion into the Jinjiang diversion area would be avoided – a practice which, without alternative flood protection measures, is becoming increasingly impractical due to rapid urban expansion and industrial development in the area.

• In the event of a 1000-year flood, water would have to be diverted into the Jinjiang diversion area, but the water level at Shashi would not exceed 45 metres, thereby avoiding a breach in the Jinjiang Dyke.

• To achieve the flood benefits claimed for the project means that this operational schedule would have to be rigorously followed – particularly during large floods. Actual operating experience of large multipurpose reservoirs during large floods indicates that there can be substantial deviations from prescribed operating procedures which result in greatly reduced flood benefits.

There is no reason to believe that the Three Gorges Project operational design would not share these problems, specifically, because of the following factors:

• **Conflict with people living in the reservoir area**
CYJV fails to recognize the most significant operational problem which would negate many of the flood benefits claimed. Roughly half a million people are currently living in what would be the designated flood storage area between 162 metres and 182 metres above sea level. These people would be inundated when floodwater was stored in the

reservoir area above 162 metres.* In the event of a flood requiring storage above 162 metres, the dam operators would have to choose between flooding out large numbers of people living in the reservoir area or large numbers of people living downstream adjacent to the river and in overflow diversion areas. For such a flood, the CYJV flood analysis fails to demonstrate that more people downstream would be protected by the proposed flood control operation at the dam, than would be flooded in the reservoir area.

• **Conflict with power generation**

Generally speaking, the higher the level of water in the reservoir, the more power can be generated. Because power generation provides cash revenue there would be strong institutional pressure to operate the dam for maximum power generation. This could be done either by delaying emptying of the reservoir, which has to be done prior to the flood season, or by keeping the reservoir higher than required for flood control. In fact, CYJV acknowledges that if its 160-metre recommended scheme is built, the reservoir may, in future, be operated at higher levels than presently stated, thereby sacrificing flood control in order to increase power generation. Such potential changes in operation are not reflected in the analysis of benefits.

• **Conflict with downstream development**

Most of the expected flood benefits are based on the assumption that population and development would increase along the beach areas and overflow diversion areas. In general, when the area downstream of a dam becomes increasingly populated and developed, dam operators frequently disregard operating rules and fail to make required

---

*This figure does not include the additional hundreds of thousands of people who would be affected by backwater in the upstream vicinity of Chongqing.

large flow releases for flood control during major floods.

At the Three Gorges Dam, operators would be faced with the difficult choice of opening the spillway gates and flooding downstream property, as required by the design, versus allowing water levels in the reservoir to rise and waiting until the last minute to open the gates while hoping that the flood would abate. If the flood did not abate and the dam operators had delayed opening the spillway gates, they would then be forced to make large emergency releases from the reservoir which could cause a catastrophic flood downstream – negating expected flood benefits. This was not considered in the probability analysis.

- **Unrealistic assumption of perfect spillway operation**

  Actual operating experience with large flood control dams show that when large floods occur there is a substantial risk that designated overflow spillway capacity is not achieved due to operator errors or malfunctions of the spillway (e.g., the experience with Tarbela, Pakistan, or Glen Canyon, U.S.).

- **Inadequate analysis of flood levels**

  In order to design the dam for flood control operation and to calculate flood benefits associated with the Three Gorges Project, it is necessary first to estimate flood risk along the middle reach by pinpointing when and where peak flows would first overtop the dykes. For large rivers like the Yangtze, with its complex flow network, diversion areas, floodplain lakes and large tributary flows, it is a complicated task to determine how water levels during a flood would vary with space and time.

  The most accurate tool for this kind of analysis is by computer simulation using a hydrodynamic model, which is capable of simulating downstream flows and water levels for various flood scenarios throughout the river system at different times. CYJV recognizes that analysis using the hydrodynamic model would provide the most accurate results but rejects its use in favour of a more analytically crude model because the "accuracy of the input data is much beyond the scope of the present study."[3] (Presumably, the input data referred to is a detailed mapping of downstream channel dimensions.) This flood routing model predicts water levels at only 12 locations along roughly 740 kilometres of river in order to identify the times and locations where the flood flow would first overtop the dykes. Use of this simplistic model in a developed country would not be acceptable for such an important flood benefit analysis.

Generally, to validate the results of such an analysis, flows computed using the flood routing model are compared with actual observed flows along the river. But CYJV's validation is not convincing because there is a discrepancy between observed and computed flows during the peak flow period of a magnitude about the same as the total flood storage capacity of the Three Gorges reservoir. CYJV acknowledges that the main reasons for this discrepancy are hydrodynamic effects, such as backwater effects and channel conditions, which it has declined to model, but nevertheless concludes that "the model represents reasonably well" the river flows.[4]

CYJV emphasizes the importance of accurate water level predictions by noting that a 20 centimetre change in predicted water level against a dyke could mean a 20 percent change in flood benefits. But the use of a flood routing model instead of a hydrodynamic model means that there are likely to be substantial errors in predictions as well as in flood control operational design. In addition, because key hydrodynamic factors affecting flood risk have been ignored, expected flood benefits achieved by upstream storage would not be as large if all these other factors had been considered. It is also possible that because the model does not predict where the

peak flows are likely to first overtop the dykes, the operational
design could increase flood risk in some areas.

Flood Benefits Exaggerated

The inadequate flood analysis and unrealistic operating
assumptions together tend to overstate the flood benefits of
the project. In addition, CYJV makes two other major
assumptions that greatly exaggerate the value of the flood
benefits:

- **Economic growth rates that increase property
  value between 10 and 30 times present values**

  Some flood protection agencies in developed countries
  consider the use of any predictions of future growth in flood
  benefit analysis as conjectural and invalid. CYJV's
  assumption that such a drastic increase in investment
  would occur over the next 60 years for land that would
  remain flood prone is highly questionable.
- **Use of the same economic growth rates with and
  without the dam**

  One of the basic tenets of flood management planning is
  that the perception of flood damage potential and the type
  and value of land use are closely related. Obviously people
  prefer not to build in flood-prone areas and, even if they do,
  they would want to flood-proof their property. These actions
  would greatly reduce the flood damage potential in the "no
  dam" scenario and hence the net benefits claimed for the
  project. If, for example, the growth rate without the project
  was 40% of that with the project, total flood benefits would
  be zero.

Both these assumptions illustrate a fundamental planning
error that permeates CYJV's flood control analysis. It focuses
on flood control storage as a goal in itself, whereas it is merely

one tool for managing floods. Rather, CYJV should have recognized the primary goal as flood damage reduction, which is achieved through an integrated strategy that includes dykes, diversion, zoning, and flood proofing.

Because of this, the project designers have built a methodological flaw into their analysis – they fail to analyze flood management of the Yangtze River as a complete system, which would require incorporating not only "plumbing" but also hydrologic, geomorphic, social and economic factors.

Project Costs Ignored
• Relocating people out of the reservoir area

If any realistic flood control operation were to be carried out, the half-million people living between 162 and 182 metres elevation in the reservoir region would have to be relocated, which could increase the total project budget by at least 20%.

• Backwater flood damages upstream

CYJV predicts that the 100-year flood level would rise at least one to two metres within 30 years due to sedimentation in the backwater near the city of Chongqing. Although the increased level of flooding would affect hundreds of thousands of people, CYJV failed to consider this in its analysis of flood damages. Rather, it defines a "critical" flood at a level of 198 metres or higher, which corresponds to the elevation of the Chongqing-Chengdu railway. This "critical" level is inconsistent with CYJV's own figures, which show damaging flows to occur at much lower flood levels.*

CYJV circumvents the problem of sediment deposition by claiming that "future upstream reservoirs and conservation

---

*In 1981, Chongqing was hit by a flood that cost $5.4 million in flood damages and affected approximately 137,000 people. The flood stage reached 194 metres elevation above sea level which is four metres below CYJV's "critical" flood stage of 198 metres.

measures" would reduce the high sediment load carried by the river during floods.[5] Neither the cost of reducing sediment input to the reservoir nor the cost of increased flood damages in Chongqing are included in project costs.

• **Repairing downstream dykes**

CYJV acknowledges that clearwater flows* would erode the downstream dykes in the middle reaches and claims this problem would be managed by river training works and dredging. Neither of these measures would prevent undercutting of the dykes downstream and it is worth noting that elsewhere in the report, when discussing alternative flood protection improvements, CYJV lists dredging as "not very promising because it interferes with the natural equilibrium of the river, and would require frequent and costly maintenance."[6] Such costs, which CYJV recognizes can be exorbitant, are not included in the cost-benefit analysis.**

• **Flood damages due to coastal erosion**

Adjacent to the Yangtze River mouth, extensive offshore mud flats are formed by the sediment discharged from the river in a cycle of sediment deposition and erosion. Several hundred kilometres of low-lying shoreline depend on the mud flats for protection against coastal flooding. CYJV failed to analyze the shoreline erosion and coastal flooding that will result from both the reduction in sediment delivery due to the effect of reduced flood peaks and the capture of sediment in the reservoir – a serious omission considering that the Yangtze River ranks fifth in the world in terms of its sediment discharge to the ocean. The costs of either increased coastal flooding or additional shoreline protection

---

*Clearwater flows: when the river's sediment is trapped by the reservoir, the river carries less sediment downstream of the dam.
**As an example, the estimated cost (as of 1980) for an ongoing program of dyke improvements is $1.8 billion.

are not included in the cost-benefit analysis.

Failure to Consider Cost-Effective Alternatives

Because the primary purpose of the project has been defined as flood control rather than flood damage reduction, the importance of other components of the flood management system have been downplayed. For example, the existing system of dykes and diversion areas provides substantially greater flood benefits than flood control storage upstream on Yangtze tributaries. Combined with flood retention in the floodplain lakes, the total flood storage capacity downstream of the Three Gorges appears to be about 200 cubic kilometres, compared to the Three Gorges reservoir's maximum flood storage capacity of 31 cubic kilometres.

CYJV states that a long-term program of flood control improvements is already under way but does not present this as an alternative nor discuss how the benefits of small improvements to the dykes would affect the cost-benefit ratio. On the one hand, CYJV states without substantiation that such improvements "would be uneconomic and impracticable,"[7] and, on the other, its analysis shows that a 20-centimetre increase in the height of dykes could increase flood benefits by 20%.[8]

One important alternative component of an improved flood management system (which was relegated to an appendix of the study and not even mentioned in the study's summary of alternatives) is the provision of ring dykes and refuge centres to protect populated areas within the diversion and beach areas. Existing ring dykes average eight metres high and encompass some 21 towns, protecting 93,000 people in the Jinjiang Diversion Area. Also, there are a number of refuge centres protected by ring dykes in this area that would serve 150,000 people in the event of flooding. CYJV concludes that "protective dykes could be justified if the annual probability of flooding is more than 1.5%," which is equivalent to a 75-year flood.[9] Since this is the case for the three diversion and beach areas which already provide most of the project's claimed flood benefits, then this would be a more cost-effective alternative to building the Three Gorges Project.

Flood reduction strategies could achieve flood benefits equivalent to or higher than those expected with the Three Gorges Project by a combination of upgrading critical dykes, modifying overflow areas, providing refuge centres and ring dykes, flood-proofing, flood-warning systems, and development zoning. Non-structural measures such as development restrictions in flood hazard areas can greatly reduce flood damages. CYJV considers development restrictions along the river proposed by the Chinese government as a strategy to minimize flood damages caused by the Three Gorges Project in the reservoir region:

> This restriction would prohibit future construction of new buildings, whether public, commercial, or residential, and new industries around the reservoir perimeter below El 182 [182 metre elevation mark]. This restriction would also effectively limit future population growth below El 182.[10]

But CYJV did not consider development restrictions as a potential strategy downstream. Using CYJV's own figures, a restriction on development in the beach and diversion areas to less than 40% of the growth predicted in the major protected areas (such as the city of Wuhan) would equal all flood benefits expected with the project.

Failure to Consider Costs of Potential Catastrophic Failure

CYJV implicitly claims that the Three Gorges Project

would prevent major loss of life in an extreme flood such as the one-in-1000-years category. There is repeated emphasis that failure of the Jinjiang Dyke would have catastrophic consequences. So, the flood control operation of the reservoir is directed towards managing water levels to prevent failure in the downstream channel, even though the probability of flood damage is much larger in other areas – for example, in the beach areas and diversion areas.

A systematic impact analysis of the project would probably indicate that the benefits of preventing loss of life are likely to be negated because the dam itself would increase the potential for loss of life as follows:

- **Increased risk of failure of downstream dykes**
  CYJV acknowledges that there would probably be an increased incidence of dyke failure due to a downcutting of the river channel which would undermine the river's banks. This vitally important impact has been dismissed without substantiation by CYJV, with their statement that "channel morphology should not change significantly because the dykes that presently border the river are in many cases protected by rip-rap."[11]* As well, undercutting and erosion would be significantly aggravated by the wide, daily fluctuations in flow because of power generation demands. Equally important as the physical impact of the project would be the institutional impact, which could increase the risk of dyke failure. Construction of the dam would draw on funds allocated for upgrading and maintaining dykes. As experience with other large flood control dams has shown, the mistaken perception of improved flood protection due to

---

*Rip-rap is the term used for large rock which is commonly used to protect the surface of earthen dykes, embankments, and river banks.

the dam could lead to reduced maintenance and deterioration of downstream dykes and other flood management infrastructure.

- **Potential failure of the dam**

  As is discussed in the later chapter on dam safety, the risk of catastrophic failure at the Three Gorges Project is probably of the same order of magnitude as the probability of a 1000-year flood. CYJV failed to prepare a map of the area which would be flooded if the dam were to fail, even though the potential loss of life would number in the millions. Property damage would be so extensive that even if the probability of failure were 1 in 10,000 years, any flood benefits claimed for the dam would be negated, according to CYJV's own figures.

- **Large increases in population induced to settle in flood-prone areas**

  Although CYJV recognizes that the presence of dykes can create a false sense of security leading to more fatalities when they fail, it has not applied this same relationship to the presence of the dam itself. CYJV's analysis of flood benefits is clearly based on the assumption that an additional population would settle in flood-prone areas. What CYJV does not discuss is that the flood risk in these areas would increase as a result of any of the operational problems described earlier. Furthermore, CYJV does not discuss how risks would increase over time as the reservoir becomes clogged with sediment and its capacity to store flood waters is reduced.

  It is reasonable to assume that, as sedimentation increases, the dam would be operated to maximize protection against smaller floods. This would lead to a scenario where cities and towns have been built up in what were thought to be protected areas and ring dykes and refuge centres have long been neglected. In the event of a large flood, the dam would no longer be able to control flood waters as originally intended, and, inevitably, the loss of life in such circumstances would be greater than if the dam had never been built.

.

Chapter Nine
Missing Energy Perspectives

by Vaclav Smil, Ph.D.

Editors' Note: As part of China's strategy to sustain economic growth, CYJV concludes that the Three Gorges Project is China's best option for alleviating electricity shortages in urban industrial centres as far away as Shanghai, 1100 kilometres downstream of the Three Gorges region. Vaclav Smil presents several key perspectives missing from CYJV's justification of the Three Gorges Project which explain not only why the project is a poor investment in China's energy future but also that environmentally and economically sound alternatives exist.

For more than two generations, Chinese and foreign perceptions of the need for the gargantuan Three Gorges Dam have remained surprisingly limited. "It is as if 50 years of technical and financial exploration left little time or energy for more thoughtful consideration of the wider implications for China's society and economy...questions...tend to be formulated in terms that constrain fruitful discourse on interrelated physical, social, and ecological aspects."[1] The CYJV study is simply the latest narrowly focused contribution to the seemingly interminable debate over the project.

China's Need for Electricity

All justifications of the Three Gorges Project, including the CYJV study, assume that China needs sustained and substantial increases in electricity generation for a central grid network. They anticipate that the average annual demand for electricity will grow by 8 percent until the year 2000 (doubling in less than 9 years), and by 5 percent during the first decade of the new century (doubling in 14 years). For the areas to be supplied by the Three Gorges Project – Central China, East China, and Eastern Sichuan – demand is expected to increase from 20.6 gigawatts (GW) in 1985 to about 70 GW by the year 2000 and to just over 120 GW in 2010.*

Annual electricity generation growth rates from 7 to 10 percent have been common in countries at comparable stages of economic development; indeed, these high growth rates prevailed even in many richer nations until the early 1970s. In the regions which would be supplied by the Three Gorges Project, the availability of electricity on a per capita basis is only one-sixth to one-quarter that of other developing countries such as Brazil or Mexico, and merely 1 to 2 percent of average per capita energy consumption in Canada or the United States, respectively.[2] China's national newspaper, the People's Daily,** reported in the late 1980s that China was short 30 terawatt-hours of electricity, equivalent to roughly 40 percent of the expected annual average output from the Three Gorges Project.

At first sight, there appears to be nothing questionable about these figures and they seem to provide the obvious rationale for going ahead with the dam: China is in great need of electricity. But a very different perspective unfolds by looking closely at how and where electricity is currently used. The U.S. Office of Technology Assessment reports that it is the orientation of China's industry and its inefficient use of energy which are to blame for energy shortages, rather than a lack of energy supply.[3]

China's heavy industries, such as iron and steel manufacturing,

---

* 1 megawatt (MW) = 10^3 kilowatts
  1 gigawatt (GW) = 10^6 kilowatts
  1 terawatt (TW) = 10^9 kilowatts
**Remrinribao (People's Daily) is the official newspaper of the Chinese Communist Party.

and chemical processing, consume about 65 percent of the country's electricity supply. In 1987, China's steel industry consumed roughly two times more electricity to produce one tonne of steel than western and Japanese steel producers. Similarly, for the production of sulphuric acid (a basic compound for many chemical processes), and ammonia (used to synthesize urea, China's leading nitrogen fertilizer), Chinese plants consume anywhere from three to six times the amount of electricity used in western countries.[4] Combined with China's generally low efficiencies for fuel combustion and electricity generation, this makes China one of the world's most energy-intensive economies.

China consumes seven times as much energy as Japan to produce one unit of gross national product (GNP).* Even compared to other poor, populous nations such as India, Indonesia, and Brazil, China's heavy industry accounts for over half of the country's GNP,** which is an inordinately high share.[5]

China's orientation towards heavy industry means that most of the country's energy is being wasted. The bulk of China's heavy industry and technology are of the 1950s vintage and are under the control of cumbersome government bureaucracies. As a result, they are inadequately maintained and operated, and since more than nine-tenths of all electricity is sold by the state at artificially low prices, there is little incentive to upgrade systems and improve their performance. The Three Gorges Project, which is intended to provide more

---

*Assuming the official exchange rate. If the exchange rate were adjusted according to purchasing power parity, China's energy consumption per unit of GNP would only be three times as much as Japan. In 1987, Japan's GNP was $1,760 billion and China's $350 billion; their energy use was 380 and 580 million tonnes of oil equivalent, respectively.
**Total industrial output accounts for 52 percent of the country's GNP.

state-subsidized electricity to the inefficient industrial sector, would not eliminate China's energy shortages, but would perpetuate existing wasteful practices.

Because so much energy is currently wasted in China, the immediate potential for improving and expanding energy services, without building new generating plants (coal, nuclear, hydro), is vast. In fact, the electricity now required by the Chinese economy – 80 percent of which is devoured by the industrial sector – could be delivered using just 60 percent of the country's existing hydroelectric generating capacity. This amounts to electricity savings of 270 terawatt-hours or nearly four times the output at Three Gorges. Another, more conservative estimate of improvements in industrial energy efficiency over two decades – the time which would be required to put the Three Gorges Project into full operation – amounts to an annual saving of at least 70 terawatt-hours, or the equivalent expected annual output from the Three Gorges Project which could be used elsewhere.

Rather than building the Three Gorges Project, a gradual shift in China's industrial structure and a widespread improvement in energy use is achievable by a combination of off-the-shelf technologies, improved system management and price reforms. Critics of the Three Gorges Project within China agree that, in an overwhelming number of instances, these changes would be less capital intensive than building new generating plants. Such a strategy would have the added advantage of providing more employment and improving the competitiveness of China's industries without creating costly environmental impacts, such as those anticipated at Three Gorges.[6] In fact, an increasing number of energy experts agree that economic prosperity increases with the greater availability of *useful* energy services rather than with the greater supply of fuels and electricity used wastefully.

Another energy-saving strategy would be to expand light

industries, such as food processing, textile manufacturing and consumer electronics, which would require only one-quarter the amount of energy used by heavy industry[7] to produce an equivalent value of industrial output. If light industries increased their share of total industrial output by just 7 percent, 30 terawatt-hours of electricity could be saved each year, or about 40 percent of the expected annual output at Three Gorges.*

China's Hydroelectricity

China is endowed with the world's largest untapped hydropower potential and therefore hydroelectricity will undoubtedly play a major role in the nation's future electricity supplies. Detailed surveys completed in the early 1980s by China's Ministry of Water Resources and Electric Power show that it would be technically feasible to build 11,103 hydro dams** with a total capacity equivalent to 27 times the planned output from the Three Gorges Project. The Yangtze's tributaries alone could potentially support 4440 hydroelectric stations with a total installed capacity six times that of the Three Gorges Project.

Since less than 10 percent of this capacity has been tapped, China is in a position to decide prudently – on the basis of social, environmental and economic factors – whether or not to build large dams at all, or small dams.***

Sichuan province, the region which would have to bear most of the burden of resettlement, has about one-quarter of

---

*Assuming all other factors remained equal.
**This figure includes dams with relatively small plants of 500 kilowatt capacity to giant plants with several gigawatt capacity.
***Megaprojects worldwide are known for their huge cost overruns and construction delays. CYJV uses Canadian experiences to estimate what the Three Gorges Project costs would be twenty years from now.

China's total hydropower potential, and could build a series of 1 to 3 GW power plants instead of the Three Gorges Project. These dams could be built on upper, less densely populated reaches of Yangtze tributaries, in less than half the time required to build the Three Gorges Project, with less social and environmental impacts. The province, and hence the country, would benefit from the new supply much more rapidly, and could rely, to a greater extent, on domestic engineering capability.

Most importantly, the development of hydropower should be undertaken only if the project poses no long-term environmental risks on a scale that is commensurate with the country's technical and investment capabilities, and only if the project meets the electricity needs of China's predominantly rural population.

Rural Energy Crisis

The Three Gorges Project is part of China's energy policy which stresses large centralized energy supply projects for electricity generation. Such investments neglect the basic energy needs of nearly three-quarters of China's population living in villages and small towns. Almost half of the rural population is without electricity; roughly half of rural households experience severe shortages of biomass fuel (straw, wood, shrubs, and grasses) for everyday cooking.*

To date, widespread construction of small hydrostations has been the most successful means of expanding China's

---

*The World Resources Institute reports that expanding biomass resources is essential for meeting basic energy needs in rural areas. Promoting the efficient use of biomass fuels would also help reduce deforestation, desertification and soil erosion. Modernization of bioenergy would not only help China reduce its reliance on coal but would also generate employment and stimulate domestic technological development.

rural electricity supply since most of these locales could never be linked to centralized electrical grids from giant dams, such as the proposed Three Gorges Project, because of high cost or difficult access. More than 70,000 rural stations with a total capacity of nearly 10 GW have been installed. More than three-quarters of China's 2,133 counties have small hydrostations and about one-third of these counties rely on them for most of their electricity.

Although experience has proven that small-scale hydro is not without technical and environmental problems – such as poor design and sizing, unreliable equipment, and excessive sedimentation – these problems are much more easily managed than problems arising from giant projects.

Plant Size

Since the early 1970s all the leading Western economies have come to recognize the perils of large dams. Even in Canada, such giant projects as Phase II of the James Bay Project face increasing opposition because of their inordinate demand for capital and skills, and their almost invariably negative environmental impacts. Above all, they are inherently inflexible: once the project is built, its high cost would render it too expensive not to operate, even when its negative impacts warrant mothballing the project.

Simplistic assessments of the Three Gorges Project as an energy supply option – detached from judging the economic merit of generating more state-subsidized electricity to supply the horribly inefficient industrial sector – may show economic benefits because they fail to account for the long-term impacts ranging from reservoir sedimentation to coastal erosion. Such impacts were anticipated at other large-scale dam projects inside and outside of China, and were judged to be acceptable and manageable in view of the huge benefits expected – but became intractable burdens just a few years or a decade after the dams were built.*

Conclusion

The most effective means of providing energy services for China's modernization has been obscured by the grandiose ambitions of dam builders in China, Canada, and elsewhere. A rigorous appraisal of alternatives would reveal that a better energy strategy lies in avoiding the inestimable cost of environmental degradation associated with large dams, in less reliance on heavy industry, and in the vigorous promotion of efficiency improvements and conservation measures through technical innovation and price reforms.

To complement these efforts, the development of China's hydropower should be undertaken only on a scale commensurate with the country's technical and investment capabilities, without the worrisome long-term environmental risks such as those posed by the Three Gorges Project, and in accordance with the needs of its predominantly rural population.

---

*For example, rapid sedimentation in the Sanmenxia Dam reservoir on the Yellow River forced a costly reconstruction and downgrading of the dam's original capacity. Also, the reduced sediment load in the Nile, due to construction of the Aswan Dam, has caused millions of dollars' worth of downstream channel and coastline degradation.

Chapter Ten
Dam Safety Analysis

by Philip B. Williams, Ph.D., P.E.

The consequence of failure at the Three Gorges Dam would rank as history's worst man-made disaster. More than 75 million people live downstream on an intensively culti-vated floodplain that provides much of China's food. It is therefore reasonable to expect that a key design criterion for the project is ensuring that the risk of failure is kept ex-tremely low.

Because of the limited operating experience with large dam projects of this type, and the disquieting number of safety incidents that have threatened the integrity of large dams in the last two decades, it is reasonable to expect that CYJV would use the best state-of-the-art techniques to demonstrate that the design, construction, operation and decommissioning of the project would keep the risk of failure acceptably low.

Unfortunately, CYJV does not address the safety issue either systematically or coherently. It provides no acceptable risk criteria, no mapping of the area and population at risk, no comprehensive risk assessment which identifies all the potential failure modes, and no identification of fail-safe measures. Because safety is not analyzed as a discrete topic, major failure mechanisms and combinations of failure mechanisms are ignored. There are many such possibilities; for example, a reservoir-induced earthquake that initiates new landslides close to the dam; sabotage or military action that disables spillway gates immediately before the flood season; unanticipated delays in construction leading to the overtopping and washing out of one of the cofferdams.*

Although CYJV discusses some safety issues, it makes many major assumptions and gross underestimations about the dam's design which effectively put the probable risk of dam failure greater than the risk of a 1000-year flood for which the project is designed. Examples of the flaws in CYJV's analysis are as follows:

- **Underestimation of earthquake ground accelerations**

   One of the most important structural design criteria for a dam is the estimation of ground acceleration in the event of what is termed the maximum credible earthquake (MCE). For the Three Gorges design, a 6.5 magnitude earthquake occurring on a fault 17 kilometres away is used for structural analysis. There are substantial uncertainties in the selection of the MCE and also in the prediction of ground accelerations at various distances from the fault. CYJV uses a ground acceleration factor only one-third the value that would be used in a reasonably prudent design.[1] CYJV's use of these values effectively results in the most optimistic interpretation possible of likely ground accelerations due to earthquakes.

- **Inadequate analysis of reservoir-induced seismicity**

   CYJV recognizes that the weight of the water in a large reservoir can initiate earthquakes. However, in developing ground acceleration design criteria, CYJV uses only historical records of earthquakes, which means that the design accelerations selected are likely to be too low and/or would

---

*A cofferdam is a temporary dam built across the river to divert the river's flow around the dam site while the permanent dam structure is under construction. Plans include construction of three cofferdams during three stages of the Three Gorges Project construction.

occur more frequently than expected. In addition, there appears to be substantial uncertainty about the movement of the most important of these faults, since CYJV stresses the "need for careful assessment"[2] – an assessment that presumably has not yet been undertaken.

The treatment of reservoir-induced seismicity (RIS)[3] is cursory and does not acknowledge the serious potential for structural damage, property damage and loss of life downstream that could occur. CYJV assumes that RIS occurs only on faults that are presently proven active, and implies that only short lengths of long faults close to the dam site would be activated. The length of faults passing under the dam itself and the displacement that would occur if these were activated are not identified. Therefore, it appears that the dam design is based on the optimistic assumption that no movement would occur on these faults, despite the experience with RIS elsewhere. For example, the Koyna Dam in India initiated an earthquake (approximately 6.0 in magnitude) that seriously damaged the dam and killed 200 people in an area that had not previously been seismically active.

- **Inadequate analysis of structural stability**

Apart from the optimistic estimates of ground acceleration during earthquakes and the fact that potential RIS is downplayed, it is clear that there are substantial unresolved problems related to the structural design of the dam which, if satisfactorily resolved, could add hundreds of millions of dollars to the project cost.

For example, with higher, more realistic assumptions for ground acceleration, the upstream face of the dam would be subject to stresses which would almost certainly cause cracking. And while CYJV recognizes that this would occur, it did not conduct the necessary analysis of the dam to identify where cracking could occur and what design modifications are needed.

Furthermore, CYJV failed to conduct a comprehensive assessment of project operation management to analyze possible failure modes such as rupture of the dam due to fault movement underneath it, and the performance of the dam during an earthquake with prior cracking.

Other examples of how CYJV has systematically downplayed the risk of dam failure are as follows:

- **Underestimation of the risks caused by catastrophic landslides**

  In 1963, at the Vaiont Dam in Italy, a landslide in the reservoir generated a flood wave that killed 4000 people. In the Three Gorges region, major landslides occur every few years, disrupting navigation and causing property damage and loss of life. CYJV states that the Three Gorges Project would result in "no significant change in slope stability,"[4] which is highly questionable considering that wide fluctuations in reservoir levels in the Three Gorges region are highly likely to have a destabilizing effect on potential slide areas.

  Although CYJV discusses the risk of landslides in the reservoir, it did not investigate the effect of earthquakes, including those induced by the reservoir itself, on activating landslides in areas it has rated as stable. Nor did it evaluate the impact of landslide waves on spillway gates at the time of rapid drawdown* in the reservoir immediately prior to the flood season.

  Finally, CYJV did not conduct a systematic analysis of zones at risk from waves 20 to 50 metres high that could

---

*Drawdown refers to the release of water to lower the level of water in the reservoir.

result from individual landslides and could conceivably kill tens of thousands of people living near the reservoir. Therefore, it appears that the threat to people living around the reservoir and dowstream, and the threat to safe operation of the dam, have been greatly underestimated.

Apart from other optimistic assumptions, CYJV's discussion of impacts due to landslide-generated flood waves assumes that all people living in the reservoir area would be relocated above the 182-metre elevation mark, contrary to the resettlement plans stating that only people living below the 162-metre elevation mark would be relocated.

• **Underestimation of risk of spillway failure**

As the world's largest hydroelectric dam on the world's third longest river, the Three Gorges Project incorporates many experimental technological innovations. One such experiment is the construction of the world's largest submerged spillway bays. Each of the 27 spillway units has a capacity equivalent to the average flow of the Missouri River in the United States. CYJV confidently asserts "there is no reason to believe that these structures could not be successfully designed, constructed and operated," even though the discharge per unit width is "well beyond proven world experience." CYJV's confidence in the spillways is further undercut in the same paragraph with the statement: "The feasibility of such a high unit discharge should be reviewed during final design."[5]

In fact, operating experience with extremely large flows through such spillways has not been good. At the Tarbela Dam (Pakistan), and the Glen Canyon and Hoover Dams (U.S.), extremely high velocities and pressures caused cavitation* and erosion which threatened the structural

---

*Cavitation: extremely high velocities cause negative pressure which can break off pieces of the spillway's concrete surface.

integrity of the dam and necessitated serious and costly repairs. Similarly, at the Three Gorges Dam there would be a high possibility of failure.

Another questionable assumption is that the "good and homogeneous quality" of the rock immediately downstream of the dam would minimize scouring (erosion of the channel caused by the river's flow).[6] Actual operating experience with this is very limited, but there is a significant possibility that scouring could threaten the structural integrity of the dam, as nearly occurred at the Tarbela Dam. Once scouring begins it is very difficult to correct and requires continual remedial measures which can add significantly to operating costs.

- **Failure to consider downstream effects of cofferdam failure**

During the construction of the project a series of temporary cofferdams would be constructed across the river in order to divert its flow. The second and third phase cofferdams would, at best, be capable of withstanding a 100-year flood and a 200-year flood, respectively. If larger floods occur, these cofferdams could quickly wash out,* releasing a flood wave that would overwhelm the Gezhouba Dam** and continue downstream to overtop the Jingjiang Dyke, drowning hundreds of thousands of people. CYJV estimates the probability of such a catastrophe to be about 1 in 20, which should be considered an unacceptably high risk.

---

*On a much smaller scale, such a cofferdam failure occurred in 1986 at the Auburn Dam site on the American River (U.S.). The cofferdam was washed out by a flood flow, only one-fifth of the volume that would be released from the Three Gorges cofferdam, but fortunately a disaster was averted by the Folsom reservoir downstream which was able to contain the flood.

**The Gezhouba reservoir capacity is less than one-half of a cubic kilometre (1/2 km³), a mere fraction of the Three Gorges reservoir capacity.

• No provision for decommissioning of the dam

The risk of dam failure increases with its age as construction materials deteriorate, mechanical systems such as spillway outlet gates fail, and the effects of a series of problems, such as corrosion, abrasion, sedimentation, and downstream scouring, become intractable.

CYJV has calculated the costs and benefits of the project over a 50-year period (for comparison's sake, Chinese culture has developed alongside the Yangtze River over some 4000 years). Regardless of the dam's economic lifespan, CYJV should have made provision in the feasibility study for decommissioning the project in a way that would ensure the safety of those living downstream. The costs of decommissioning should have been included in the cost-benefit analsyis.

Chapter Eleven

Sedimentation Analysis

by Philip B. Williams, Ph.D., P.E.

The Yangtze is not only a river of water, it is also a river of sediment. The flow of the Yangtze carries with it the fifth-largest sediment discharge of any river in the world, equivalent to about 4 percent of all river-borne sediment discharged to all the oceans of the world.

In a river like the Yangtze, most of this sediment is conveyed during floods, and it is sediment deposited from these floodwaters that has formed the extensive fertile flood-plain of the Yangtze Valley downstream of the Three Gorges. As the river flows across the valley to the sea its waters erode and redeposit the sands and silts of its bed and banks.

This dynamic equilibrium between the river's erosion and deposition creates the slope and course of the river channel. It is on this active form of the river channel that cities have grown up and levees for flood defence have been constructed. Further downstream, the Yangtze estuary and its low-lying fertile coastal plains have been created by sediments brought down by the river, and major cities like Shanghai have been established on these fragile landforms.

When a large reservoir is filled with water, the flow of sediment through the river system is interrupted. The coarser sediment materials, the boulders, gravels and sands, are deposited at the upper end of the reservoir forming a delta. Almost all of this coarser material is carried by the flow along the river bed and is referred to as bed load. The amount of bed load moved is determined mainly by the flow velocity, with most of the bed load moved during high-velocity flood events. When the velocity of the river decreases suddenly as it merges with the slow-moving reservoir water, the bed load is deposited raising the river bed, a process known as aggradation. Because the river channel fills with sediment, floodwaters spill out more frequently onto the adjacent floodplain.

The bed load entering a large reservoir, like that of the Three Gorges, is trapped in the reservoir. The discharge, free of bed load, will erode sediment from the bed and banks downstream. This causes the lowering of the river bed, a process known as degradation. Degradation downstream of a dam the size of the Three Gorges would extend for hundreds of kilometres, threatening cities and populations that live downstream.

Usually the bed load comprises a small but important proportion of the total sediment load. Researchers typically estimate the proportion of bed load in large rivers to be in the range of 2 to 8 percent of the total sediment load. The rest of the sediment, generally consisting of fine sands, silts and mud, is referred to as the suspended load. This finer material takes longer to settle in a reservoir, and is therefore normally transported further into the reservoir than the bed load.

It is only in the last few decades that dam designers have started to recognize the need to evaluate thoroughly the processes of reservoir sedimentation and the long-term impacts on the dam and the river upstream and downstream. Now, with about 1 percent of the world's total reservoir storage capacity lost to sedimentation each year, there is a growing recognition of the need to extend the lifetime of these reservoirs as long as possible by attempting to use new types of operations that try to maximize the amount of suspended sediment flushed through the reservoir.

Techniques to deal with reservoir sedimentation problems are still in the earliest stages of development, and there are still many problems inherent in sedimentation analysis that have not been worked out. Despite these problems, the proponents of the Three Gorges Dam have put forward a dam design and method of operation that they say will deal completely and effectively with the sedimentation problem.

A review of the sedimentation analysis and the design and operation procedures for dealing with sedimentation, contained in the CYJV feasibility study, reveals fundamental errors in the analysis done by CYJV and flawed premises on which the CYJV analysis is based.

General Problems with Sedimentation Analysis

The planners of such a large-scale dam project as the Three Gorges have an extraordinarily difficult task in analyzing and predicting the impacts of sedimentation, for the following reasons:

First, the type of operation and design proposed to minimize sedimentation has not been successfully demonstrated on such a large dam project. In effect, the Three Gorges Dam is a gigantic experiment in river management.

Second, the scale of the river and sediment discharges that the planners are attempting to manage is unprecedented. CYJV reports that total sediment discharge at Yichang is 530 million tons/yr, equivalent in volume to about 0.43 cubic kilometres (km^3) per year, compared to the total normal reservoir power pool volume of 26 km^3. The only previous attempt to manage sediment flows of this magnitude was at the Sanmenxia Dam on the Yellow River, which is widely acknowledged outside China to have been a costly failure due to unanticipated sedimentation problems.

Finally, the level of confidence required of the feasibility analysis in recommending flushing as a sediment control mechanism exceeds the confidence limits of the state-of-the-art of the sciences of sediment hydraulics and fluvial geomorphology. Even with the best analytic and data collection methods, sediment discharge can only be predicted within error bands of several hundred percent for a given flood event.

Recognizing these difficulties and taking into account the importance sedimentation analysis has for critical design issues – namely, the useful life of the reservoir, navigation, and the impacts on millions of people living upstream, downstream and on the estuary – it would be reasonable to expect that CYJV's analysis would adopt a conservative approach. For example, one would expect CYJV to take into account actual worldwide experience with reservoir sedimentation control. Also CYJV should have reviewed worldwide experience with river channel changes due to reservoirs. Furthermore, it was CYJV's responsibility to test independently YVPO's data, and the sensitivity to uncertainties of this crucial analysis.

Inconsistent and Incomplete CYJV Analysis

CYJV did not review YVPO's original data nor did it analyze the long-term consequences of sedimentation in the Three Gorges reservoir. Despite this incomplete review, CYJV made a finding that is unprecedented in the field of sediment hydraulics: CYJV predicted that the Three Gorges reservoir storage can be preserved indefinitely. But, the data and analyses on which CYJV bases this extraordinary finding are inconsistent, incomplete, and fail to substantiate the claim.

CYJV states categorically in its summary document written for decision makers and potential financiers, "about 90% of [the reservoir's] effective storage can be preserved indefinitely. Reservoir sedimentation will not limit the useful life of the project."[1] However, a closer reading of CYJV's technical report on sedimentation qualifies and undercuts this sweeping statement.

According to CYJV's sediment volume, "preserved indefinitely" actually means "85 to 90% of the regulating and flood control storage will be preserved after 100 years of operation, when the system reaches or is approaching equilibrium."[2]* Equilibrium is further qualified as follows: "Approximate equilibrium is reached when 90 to 95% of the sediment entering the reservoir is flushed through the reservoir."[3] What this really means is that CYJV estimates that after 100 years, about 50 percent of the reservoir volume will have filled up with sediment,[4] but maintains that only about 10 to 15 percent of the active power pool will have been filled because CYJV predicts most of the sediment will fill the deepest part of the reservoir. Furthermore, CYJV predicts that over the first 100 years of operation, the rate of siltation in the reservoir will decrease from about 70 percent[5] of incoming sediment to only 10 percent.

Although CYJV acknowledges that sediment will continue to accumulate in the reservoir after 100 years, it does not address the inevitability of most of the reservoir storage ultimately being lost. Nor does it address the future task of decommissioning a reservoir like Three Gorges when much of its storage has been filled with sediment.

As well as being contradictory and incomplete, CYJV's sedimentation analysis is unreliable because it is based on flawed premises.

Flawed Bases of CYJV's Sedimentation Analysis

1. Incomplete Review of YVPO Analysis

CYJV's task was to make an independent review of YVPO's sedimentation predictions. CYJV concluded that this work was "very satisfactory" and "of consistently high quality."[6]

---

*Perfect equilibrium is achieved when the same quantity of sediment enters, as is flushed out of the reservoir.

CYJV came to this conclusion apparently without verifying YVPO's original data. Although CYJV lists the main sedimentation reports prepared by several Chinese agencies and research institutes, it admits that only "a number of brief summary reports and publications from these groups were made available by YVPO."[7]

While China has experienced and competent hydraulic engineers, it is not clear what role they have played in selecting the key assumptions used in YVPO's analysis. The question of feasibility of the project is highly politicized and if there was a negative conclusion to the sedimentation analysis, it would be difficult to argue that the project makes sense. In these circumstances, as an outside reviewer, CYJV has the responsibility to evaluate rigorously the key assumptions and the degree of confidence in these assumptions, and to verify the veracity of the data used.

It does not do so, but rather replicates the model developed by YVPO and uses the same assumptions and unverified sediment data. Not surprisingly it produces similar results, which it then presents as confirmation of the YVPO results.

2. Unreliable Equilibrium Slope Calculation

The premise on which CYJV appears to place most importance and which is described in the executive summary[8] is a prediction that sediment would deposit in the reservoir at an equilibrium slope. Once the sediment has settled into an equilibrium slope, the same quantity of sediment would enter and be flushed out of the reservoir (i.e., no additional sediment would be trapped behind the dam).

The estimation of the equilibrium slope is extremely important because it not only affects the estimation of reservoir sedimentation but also aggradation and navigation in the river channels upstream. If the slope prediction is underestimated by only 10 percent, the river bed and flood levels upstream in Chongqing would be about 4 metres higher, flooding out hundreds of thousands of people and impeding navigation.

CYJV predicts that such an equilibrium slope, extending from the bottom of the dam's outlet structure to the upstream end of the reservoir, would be shallow enough to preserve most of the reservoir volume indefinitely. CYJV based this assumption on several methodologies – regime equations, sediment transport equations and a numerical model.[9]

A closer examination of these methods shows all of them to be inadequate for the purposes of predicting sedimentation rates, volumes, and equilibrium slopes, to the level of confidence required for such an important design consideration.

These methods cannot predict the equilibrium slope with any reasonable confidence for the following reasons:

• Regime equations are based on empirical data from smaller rivers and irrigation canals. An authoritative review states: "Regime relations should never, of course, be applied in cases in which the flow, sediment transport, and channel characteristics differ widely,"[10] as is the case for the Yangtze.

• Predictions of slope based on the sediment transport relations are very dependent on the size of sediment selected. CYJV has calculated the equilibrium slope based only on sediment sizes between 0.1 and 1.0 millimetre (mm), assuming a median of 0.23 mm. If the amount of coarser material has been underestimated (see #3 below) the slope would be considerably steeper. For example, if the median sediment diameter was 0.26 mm the equilibrium slope could be 10 percent steeper, causing flooding in Chongqing.

• The one-dimensional model, and the other computational techniques, assume that flow through the reservoir can be represented by a single averaged value of velocity even though the reservoir is typically about 1 kilometre wide and

60 metres deep. This ignores the complex flow structure in the reservoir, the highly variable geometry (unlike a river) and the effect of three-dimensional density currents in distributing sediment throughout the reservoir.

- Equilibrium slope predictions are extremely sensitive to the hydraulic roughness selected. CYJV data shows that the observed roughness can vary by about 100 percent depending on river stage and location within the Gorges. The report points out that a 15 percent error in slope, equivalent to about an 8 percent error in roughness, would cause the bed level at Chongqing to be 6 metres higher, flooding out hundreds of thousands of people.

- CYJV does not take into account the importance of cohesion in deposits of fine sediment that limits scouring. By failing to do so, CYJV has overestimated the ability of the reservoir operators to resuspend and flush out deposited material.

- CYJV does not consider the extent to which the deposition of coarser bed load material will affect the equilibrium slope over time.

If CYJV's prediction of an equilibrium slope is not correct, sediment could continue to accumulate in the reservoir, rather than being flushed through the dam. This would decrease the storage capacity of the reservoir, cause aggradation upstream that would result in flooding and threaten hundreds of thousands of people near Chongqing, and degradation downstream that would result in erosion and threaten downstream development.

3. Miscalculation of Bed Load

CYJV has seriously underestimated the bed load of the Yangtze.

CYJV accepted YVPO's estimate that the bed load (defined as sediment larger than 1 mm) conveyed by the Yangtze is only 0.05 percent of the total sediment. This estimate is contradicted by statistics from the Yichang Hydrological Gauging Station which indicates that the Yangtze's bed load is 1.6 percent of its total sediment load.[11]

Nonetheless, CYJV's entire sedimentation analysis is based on a bed load of 0.05 percent, an amount so small that CYJV counts it as zero in the reservoir sedimentation analysis.

Even though CYJV acknowledges that "the quality and quantity of basic field data is of crucial importance to the sediment load investigation"[12] it admits it did not even review YVPO's bed load field data. Nevertheless, CYJV judged the YVPO studies based on these data "to be well done."[13]

Getting an accurate measurement of bed load is extremely difficult, even for small rivers, because of practical difficulties extracting samples from the bed of a fast-flowing river. Sediment transport rates vary across the river bed and over time, even for the same flow. Apart from experience in other rivers, there is substantial evidence that the bed load of the Yangtze is significantly greater than CYJV has determined. For example:

- The existence of extensive boulder and gravel bars throughout the Three Gorges that move downstream during floods.
- The quantity of sand and gravel to be excavated from the river bed downstream of the dam site for concrete aggregate for the Three Gorges Dam. CYJV states "there is no concern about the availability of these materials."[14]
- Reports from other Chinese researchers describe gravel and boulder deposits in the Three Gorges more than 35 metres thick.

Underestimation of the bed load would in turn cause the underestimation of the equilibrium slope of reservoir sedimentation, the rate and extent of upstream river aggradation, the difficulty in managing the river for sediment pass through, and the rate and extent of downstream river degradation, all of which have major negative implications for the technical, economic and environmental feasibility of the project.

4. Inadequate Empirical Estimate of Reservoir Trap Efficiency

The reservoir trap efficiency method used by CYJV to calculate reservoir sedimentation, although independent of the equilibrium slope method, cannot estimate sedimentation in the active reservoir storage zone (above 140 metres) and therefore cannot be used to predict its rate of filling. Moreover, this method underestimates sedimentation in the dead storage zone (below 140 metres) for the following reasons:

• It assumes constant reservoir levels during floods. Yet the purpose of flood control is to store water during floods. The larger the flood, the greater the sediment inflow and the greater the amount of water stored. In a 20-year flood, for example, the reservoir is expected to reach 160 metres, doubling its volume. This in turn, based on CYJV's method, could increase the trap efficiency from about 10 to 30 percent, thereby substantially increasing sedimentation in the reservoir.

• It ignores the considerable reservoir volume stored during large floods in the sloping water level of the reservoir. During moderate to large floods, even if the outlet level can be maintained at 140 metres, the reservoir level at the upper end may exceed 160 metres.[15]

• There is considerable scatter in the data from which these trap efficiency curves have been derived. For example, trap efficiencies for the envelope curves can range from 60 to 85 percent. CYJV selected 60 percent, characterizing the lower envelope curve for fine sediment, but without taking into account the error band of this data.

Underestimation of Reservoir Sedimentation Rates

In addition to the inadequacies of these methodologies to calculate sedimentation rates, the total sedimentation rates have been underestimated because of:

- The underestimation of bed load sedimentation. This could account for at least 1.6 percent of the total sediment load, based on estimates by the Yichang Hydrological Gauging Station, as opposed to 0.05 percent adopted by CYJV.
- The effect of landsliding into the reservoir. Within a 100 years, landsliding could fill several cubic kilometres of the reservoir. More important, where massive landslides occur, erosion-resistant control points would be formed within the reservoir, preventing the scouring of accumulated sediments upstream and resulting in induced upstream flooding.
- The potential increase in sediment delivery rates as watershed conditions deteriorate.
- The episodic nature of sediment delivery. After a large flood considerable aggradation could occur that may take decades to erode to an equilibrium condition. This places considerable doubt on whether the concept of equilibrium slope has real meaning in managing systems of this size.

Implications of Flaws in Sedimentation Analysis

CYJV has not convincingly demonstrated that its Three Gorges sedimentation analysis can realistically predict the actual performance of the reservoir, nor has it carried out a systematic sensitivity analysis of the cumulative effect of uncertainties in its predictions. If it had done so it should have concluded:

- There is a significant risk that sedimentation rates in the reservoir, similar to those observed in other major reservoirs, could substantially impair the performance of the project in its economic lifetime.
- There is a significant risk that aggradation of the river bed upstream past Chongqing will be substantially higher than estimated, flooding hundreds of thousands of people.
- There is a significant risk that the aggradation of the river bed upstream and the degradation of the river below the Gezhouba locks will greatly impede navigation. Because the bed load has been underestimated, the costs of dredging navigation channels have been underestimated.

There is no realistic way the reservoir can be managed, as it fills with sediment and loses its flood control storage, to protect millions of people who have been induced to move into flood-prone areas downstream.

The capture of sediment in the reservoir will cause significant degradation of the river bed for hundreds of kilometres downstream, eroding flood control embankments, undermining bridge crossings and changing the hydrologic regime of the river on which millions of people depend. The long-term costs of additional levee repair have been seriously underestimated in the study.

There is a significant risk that sediment captured in the reservoir would accelerate coastal erosion. (This is acknowledged in a technical appendix[16] but ignored in the main report.)

If the Three Gorges Project is completed as planned, it is probable that within a few hundred years the reservoir will almost entirely silt up, creating an unprecedented hazard to the millions of people living downstream, whose culture has survived and prospered for the past 4,000 years with wise management of the Yangtze River.

Chapter Twelve
Economic and Financial Aspects

by Vijay Paranjpye, Ph.D.

The feasibility study of the Three Gorges Project was conducted by the CIPM Yangtze Joint Venture (CYJV) with the principal objective of providing impartial technical input to the Government of China in its decision-making process, and to provide the basis for securing funding from international financing institutions. In the study summary, CYJV states its objective as:

> To establish firmly whether the Three Gorges Project is technically, economically and financially feasible on a basis acceptable to international financing institutions.[1]

The report was to be comprehensive with respect to costs and benefits related to flood control, power generation, navigation, resettlement and environment. Further, the study was supposed to identify the least cost option from a range of four schemes defined by normal pool levels (NPLs) at 150, 160, 170, and 180 metres elevation.

CYJV used the technique of cost-benefit analysis to arrive at the conclusion that for a total economic cost of $3.7 billion,* spread over 18 years, there would be a net benefit-cost ratio of 1.48. Based on this analysis, CYJV states:

---

*The economic cost is the figure used in the cost-benefit analysis after adjustments have been made for financial factors such as taxes and subsidies. Financial costs are the expected costs to the project builder for all project inputs. The financial cost of the project is $6.6 billion discounted to mid-1987 prices, and $10.7 billion including cost escalation.

> The Three Gorges Water Control Project is
> an attractive solution to reduce the flooding
> and improve navigation on the Yangtze, and
> will be a new major source of renewable
> energy.[2]

The detailed and sophisticated analysis of the cost and benefit values appears to be quite convincing and logical, but as one starts looking more closely at the underlying assumptions, and at the data on which the analysis is based, the shortcomings and discrepancies become apparent.

The shortcomings and flaws in the feasibility study are of two types: those which are rooted in the conceptual framework of the cost-benefit analysis, and those which are errors of omission and of commission in the process of cost and benefit computations.

The Absence of River Basin Planning on the Yangtze River

One would expect that for the largest river in China, planning a project on the scale of the Three Gorges Dam would take place within the context of a systematic analysis of the entire river basin. Strangely enough, the study makes no reference to any such analysis even though hydroelectric dams are hydrologically interdependent; the design and operation of one dam directly affects that of other projects in the river basin.

The lack of river basin analysis is even more surprising in view of the fact that the main river channel of the Yangtze is joined by more than 700 sizeable tributaries; each having development potential for hydropower, irrigation, and flood management, and each project having a different impact on the short-term and long-term development of this magnificent river.

The CYJV economic and financial analysis begins with the statement:

> The Three Gorges Project will be the only economical way to significantly increase flood protection in the middle reaches of the Yangtze.[3]

However, the CYJV report contains no analysis to support this assumption. As well, CYJV states that the feasibility study:

> Demonstrates that the project represents the lowest cost solution for the benefits obtained and explains how project features can be optimized within a range of alternatives.[4]

But the so-called alternatives are four different reservoir operating levels which are intra-project variations and therefore, in the context of the river basin, they are not real alternatives at all.

The study thus provides a biased technical input which does not arrive at a comprehensive "optimum solution," but gives a contrived justification for securing funding from international institutions.

Assumptions Underlying the Cost-Benefit Analysis

It is normally assumed that there is a *market* price for all the cost and benefit items to be enumerated and analyzed. Further, that all the perceived gains and losses can be quantified in *economic* prices or *shadow* prices (i.e., after adjusting the market price by deducting the tax elements and adding the subsidy components). In the case of values which are perceived to be significant but carry no market price, or are not traded, a serious attempt is usually made to construct models of surrogate markets in which shadow prices may be derived.

In the case of projects like the Three Gorges Project, this procedure is made difficult because the two major benefits, flood control and hydropower, do not have competitive market prices in China. Let us start by examining the benefits of flood control, which actually translate into economic losses avoided. Losses due to a major flood are caused not only by the actual physical destruction due to inundation, but also depend on the *pre* and *post* flood circumstances (or disaster preparedness). The pre-flood warning system and the evacuation system, for example, would substantially change the total value of losses. A repeat of the 1931 flood would be far less serious today because of a vastly superior advance warning and evacuation system. Similarly, the aftermath of such a flood today would be much less severe due to a more efficient public health system and faster means of goods and food transport.

It is important to note that the CYJV analysis contains a bias that leads to an overestimation of the flood control benefits of the Three Gorges Dam. First, the CYJV analysis focuses mainly on losses subsequent to the future rate of economic growth. They estimate flood damages under a high economic growth scenario to be $2.9 billion, and $1.6 billion for a low economic growth scenario. Meanwhile, CYJV largely neglects to estimate the effect of *pre* and *post* flood circumstances – both of which will significantly affect the amount of flood damage. A natural calamity of an equal magnitude in future, would likely cause less damage than that recorded in the past.

It must be stressed here that the objection is not to making investments for flood control, but to the arbitrary quantification of *losses avoided*, with the intention of justifying a particular investment project.

Next in importance to flood control are the benefits that would accrue from the generation of hydroelectricity. To begin with, let us acknowledge the fact that electricity in China carries an *administered monopoly price*. CYJV uses 1.4 cents per kilowatt-hour as the price of electricity, but this figure is actually the average revenue obtained from Central China's power grid and is lower than the marginal cost of generating power.

Therefore, the present rate of growth in the demand for electricity, and consequently the CYJV demand forecast of 122,800 megawatts by the year 2010, is based on a low price situation and is therefore accompanied by a highly inefficient end-use of electricity, particularly within the industrial sector. In all likelihood, if the price were higher, the end-use of electricity would become more efficient and the demand would be lower than CYJV's forecast – a scenario which is quite possible as China's economy undergoes progressive decentralization. But CYJV chooses to ignore this, stating:

> Whether increases will have a dampening effect on the future demand for electricity is unknown and has not been accounted for in the forecasts.[5]

CYJV does not incorporate this scenario into its calculations of benefits from power or in the calculation of the financial rate of return. The objective of calculating the financial rate of return is to determine whether the authorities could afford to borrow money for the project at commercial rates of interest. CYJV works out a financial rate of return of 8.9 percent assuming the price of power to be 1.4 cents per kilowatt-hour. It also claims that this rate could easily be increased to 1.7 cents per kilowatt-hour or 2.2 cents per kilowatt-hour, in which case the financial rate of return would be 10.16 percent or 12.18 percent. Here it is necessary to point out that in China, increasing the price per unit of power can be done just by a stroke of the pen. It would therefore be possible to demonstrate the financial feasibility of the Three Gorges Dam even if it were to cost twice as much. Similarly by extension, increasing the height of the normal pool level, thereby displacing more people and raising re-settlement costs, could always be justified, financially, by continuing to increase the price of power.

It is surprising that the CYJV experts should feel convinced that the Chinese project authorities would not deviate from the recommended normal pool level of 160 metres, especially when the dam height of 185 metres easily permits such changes. Unfortunately, what appears to be a rational benefit-optimizing exercise on the computer screen, could easily be converted into a horrendous nightmare for the displaced population. The experts working on computer terminals often tend to slur over the fact that in countries like China and India, human lives and their welfare carry a very low premium when this interferes with the execution of centralized development megaprojects.[6]

The Discount Rate

In theory, the rate of discount reflects the social cost of capital invested, or the opportunity cost of capital. The discount rate also implies the degree of importance society ascribes to a benefit which will accrue in the future, as compared to accruing that benefit today.

In China, the standard discount rate used by the Ministry of Water Resources and Electric Power is 10 percent, which CYJV applies to the expected costs and benefits over a 62-year period. The World Bank, one of the potential financiers for this project, applies a 12 percent rate of discount in its economic appraisals (as it did for the Narmada dams in India). The CYJV economic feasibility study should have used the 12 percent rate of discount; why it did not do so becomes obvious when we see, as CYJV determined, that:

> Applying a 12% discount rate causes a 15% decline in costs and almost 30% decline in benefits. The net project benefits decline by 59%.[7]

The choice of discount rate is therefore biased.

The Rate of Exchange

CYJV uses the administered rate of exchange, 3.7 yuan per U.S. dollar, in its economic analysis of the Three Gorges Project. However, economic analysts generally agree that the current rate of exchange between the yuan and the dollar is highly unrealistic, and, like the rouble in the Soviet Union, the Chinese yuan is highly overvalued. CYJV acknowledges this:

> Indications that the yuan is overvalued come from several sources. Chinese foreign exchange reserves have fallen quickly since late 1986, indicating a need to make imports more expensive in order to conserve foreign exchange. Based on these indications, it has been suggested that a rate of 5 or 6 yuan per U.S. dollar may be more appropriate than the existing 3.7 yuan per U.S. dollar.[8]

CYJV also recognizes that:

> A drastic drop in the exchange rate to about

6 yuan per U.S. dollar would increase dis-
counted construction costs by 30%.[9]

Recent changes in China's economic policy, especially
since the 7th Five Year Plan (1986 – 1990), suggest that such
a devaluation of the yuan is imminent.

The foreign import component of the project cost is presently estimated to be from 12 to 18 percent, and so, if the free
market exchange rate is applied, instead of the administered
rate of 3.7 yuan per U.S. dollar, the financial as well as the
economic cost would rise substantially, thereby making the
project less attractive. The use of the administered rate of
exchange of 3.7 yuan per U.S. dollar to estimate base costs of
the project is therefore incorrect, biased, and results in an
underestimation of the total cost in economic terms.

Time and Cost Overruns

Past experience in China and around the world has shown
that megaprojects are rarely completed on schedule. Shortage of funds, bottlenecks in management and phases of
construction, multiplicity of decision-making bodies, and
technical problems, all contribute to project delays. Of course,
the principal cause of delay is cost escalation, which can be as
much as 100 percent or more.

The twin problems haunting designers and planners of
megaprojects are cost and time overruns which fuel each
other and are directly related: the greater the cost overrun,
the greater the time overrun, and vice versa. The Gezhouba
Dam downstream of the Three Gorges Project site is a case in
point; its cost overrun was estimated at over twice the initial
budget. In China, 2 to 5-year delays in construction are quite
normal, a factor which would make the Three Gorges Project
totally unattractive, economically.

The CYJV sensitivity analysis acknowledges that a delay

in the Three Gorges construction schedule by even one year (therefore causing a delay in the commencement of power generation) would reduce net benefits by $460 million or 22.5 percent. Because of this, CYJV states that "considerable additional construction expenditures can be justified to maintain the schedule."[10]

Generally, delays at other dams have caused funds to run short, diverting funds allocated for resettlement and other development activities to dam construction. Project authorities usually justify such a diversion of funds in the name of national interest, while, in actual fact, it amounts to involuntary patriotism and compulsory sacrifice.

Estimation, Underestimation, and Omission of Project Costs

A critical assessment of CYJV's cost estimation procedure is important because it would indicate the proportion of tax or subsidy element contained in the financial price. It would also, therefore, expose any bias in CYJV's cost estimates. Unfortunately, CYJV deleted a lot of important information ("pursuant to 20(1) b,c,d, Access to Information Act"[11]) from the feasibility study before its release, which makes it impossible to determine whether the CYJV economic costs (as against financial) are correct or not. For example CYJV's costing of civil and mechanical works, project management costs, commissioning costs/rates, wage rates, labour costs, and cost of fuel, etc., have been deleted from the study. Aggregate values do not mean much if the rates per unit, wage rates of different categories, etc., are not known. It is precisely in such a situation that bias and arbitrariness creep in.

Underestimation of Resettlement and Environmental Mitigation Costs

CYJV assumes, arbitrarily, that environmental studies

and environmental mitigation measures would cost "2% of the construction and resettlement costs"[12] despite the report's admission that "detailed cash flows for the environmental costs have not been prepared."[13] In other words, the cost of environmental damage is assumed to be fixed as a proportion of total costs, even though it is widely accepted that environmental damage has increasing diseconomies of scale.

For example, if a larger dam is built, more involuntary resettlement in upstream areas would be necessary. This would cause changes in land use that would accelerate the rate of deforestation, erosion, and general deterioration of the habitat. This would increase the rate of sedimentation in the reservoir and reduce the economic life of the dam. The problem of sedimentation could then be mitigated in a number of ways: by constructing another dam (or dams) upstream mainly for trapping sediments – which would only be a short-term solution, and the economic viability of such a dam project would also have to be demonstrated; by dredging large masses of sediment which are expected to settle in the river channel near the port of Chongqing; by carrying out massive rehabilitation of the upstream catchment area which would not only conserve the soil, reduce the rate of water runoff, and reduce the rate of soil erosion, but would also significantly increase biomass fuel production.

Theoretically, the cost-benefit analysis requires that all costs necessary and essential for the accrual of project benefits must be included in the analysis. Therefore, expenditure to reduce sediment input to the reservoir should form a legitimate part of the project cost, but CYJV did not include such costs. Nor did CYJV anticipate exponentially increasing environmental mitigation costs with increasing scale of dam. Nor did CYJV collect the primary data necessary to estimate the costs associated with possible downstream farming and fishing losses, bank erosion and channel shifting.

CYJV also failed to include the cost of disruption of navigation downstream of Chongqing during the dam's construction period. In addition, CYJV reports that the Ministry of Communication may ask for $270 million compensation but did not include this in the total cost estimate, whereas the benefits from navigation after project completion have been calculated in elaborate detail.

Faulty Selection of the CYJV Recommended Project

Forced resettlement is always traumatic for those people forced to move, irrespective of the quality of the resettlement plans on paper. Because of this, planners and designers of megaprojects usually try their utmost to minimize the number of people to be involuntarily resettled. But the CYJV analysis, despite its professed objectivity and social sensitivity, has recommended a reservoir height which maximizes net benefits rather than minimizing the number of people to be forcibly resettled, as shown in Table 1. The figures in Table 1 are based on CYJV's own data and do not appear in a consolidated form anywhere in the CYJV study.[14]

Table 1 – Net Benefits versus Resettlement Costs		
	Option A	Option B
Normal Pool Level	150	160
Flood Control Level	130	140
Net Benefits	1,624	1,760
Resettlement Costs	1,687	2,170
Benefit Cost Ratio	1.49	1.48
Population	539,000	727,000
*Costs and benefits in US$ millions		

In Table 1, option A is overwhelmingly superior to option B in all respects except for net benefits. Option A has a superior benefit-cost ratio which means a better return per dollar invested. The project submerges a far smaller area, avoids the forced resettlement of 188,000 persons, and costs almost $540 million less. If any group of economists or development planners were shown these two options, almost all would plumb for option A. But in spite of the overwhelming superiority of option A, the experts of CYJV have recommended option B. This proves, quite conclusively, that the engineering consultants' attitude towards resettlement is downright callous.

Further, CYJV wrongly assumes that the cost of displacement per person remains constant, irrespective of the scale or nature of displacement. CYJV takes the estimates prepared by the Yangtze Valley Planning Office for a 175-metre-high dam with alternative reservoir operating levels, and extrapolates them for the 185-metre-high dam with alternative reservoir operating levels. To begin with, the Yangtze Valley Planning Office calculation is an underestimation because many of the non-market values lost by the displaced persons are left out and the compensation is not based on 'full replacement' cost of the assets lost.

Secondly, anyone familiar with the rehabilitation process of people and communities who have no occupational mobility or professional expertise knows that an increase in the scale of displacement causes an increase in per capita resettlement cost. Not only do the overheads increase, but the external diseconomies of scale also increase rapidly. For example, when a large number of persons are settled involuntarily in an area where there is a scarcity of life support resources – land, water, fodder, and fuelwood – the host population look upon the new settlers as encroachers, leading to a serious clash of interests, and social disharmony. In economic terms, the increase in the scale of displacement leads to a greater than proportionate increase in internal resettlement costs, as well as the externalized costs of social disruption and environmental degradation. For these reasons, CYJV's extrapolation of resettlement costs, from the 175-metre dam to the 185-metre dam involves a major underestimation.

Another fundamental reason for the increasing diseconomies of scale is that the most productive lands in the river basin are located at the bottom of the valleys and will be submerged by the reservoir. The higher land, which would be used for resettlement, becomes progressively inferior with higher elevations. The more people that are displaced, the higher up and poorer the land on which they must be resettled.

The CYJV estimates also neglect the following factors which would significantly increase the costs of resettlement:

- The impact of sedimentation in the reservoir and the consequent rise in river levels near the city of Chongqing, requiring still further resettlement.
- The existence of thousands of "non-persons" living in the city, without government permission, who would be displaced without compensation.
- The uncertainties in the estimates of the unregistered population, the natural population growth rate, and migration to the reservoir region.
- The risk and uncertainty analysis in the CYJV study indicates that the variations in numbers of people who deserve compensation, replacement land, jobs, housing, and so on, could result in a total cost increase by as much as 29 percent above CYJV's base estimates.

If all of the above costs were included in the cost-benefit analysis, it may well make the project non-viable even without unfavourable changes in the future. If further legitimate costs, such as transmission and distribution losses, which can be as high as 10 to 12 percent of the total amount of electricity generated, are deducted from the power benefits, the project would become even more unattractive.

The Impact of Inflation

The cost-benefit analysis is totally incapable of handling the impact of a general price inflation. The World Bank has estimated the annual price inflation in China to be about 4.5 percent from 1990 to 1995. Typically, economists wish it away by assuming that inflation affects the costs and benefits values equally, and therefore, the relationship between costs and benefits will hold good so long as the analysis is conducted at constant prices. In the case of the Three Gorges Project, the CYJV analysis is conducted with reference to mid-1987 prices. Of course, all analysts are aware of the fact that some prices escalate faster than others, causing relative price differentials. In China, as in many other countries, project construction costs (economic) tend to rise faster than the price of benefits such as the price of power per unit, or the price of agricultural products. This has a tendency to reduce net benefits thereby lowering the ratio of benefits to costs.

The Lack of *Ex Post Facto* Economic Evaluations

The best procedure for verifying the CYJV base-case assumptions regarding anticipated cost, benefits, project construction schedules, distributional impact, environmental impact and chances of mitigation, success of resettlement and rehabilitation plans, rates of sedimentation, etc., is to analyze past performance and experience with similar projects. But unfortunately, the religious fervour with which the dam project authorities have conducted *ex ante* studies has been totally lost once the dam has been built. Therefore, one can find very few, if any, studies which have elaborately checked out the reliability of such 'base-case' assumptions.

The next best procedure, albeit very unreliable, is the risk and sensitivity analysis which CYJV conducted diligently. Yet in the absence of ground-checks in the form of *post facto* studies, the risk and sensitivity analysis remains, at best, in the realm of speculation. Certainly such analysis should not guide important investment policy decisions.

Financial Capability

At current prices the project is estimated to cost about $10.7 billion of which about $1.4 billion are foreign costs. But the total figure could easily go up to $13.5 billion if all known costs are included.

In any case, China does not appear to have the financial capability to contribute the necessary funds. CYJV did not consider China's shortage of funds as an uncertainty in its risk analysis, and the failure to do so is a major omission.

Macro-economic Impact and Regional Distribution of Benefits and Costs

One of the serious flaws in the conceptual framework for the CYJV cost-benefit analysis is its preoccupation with the question: What benefit at what cost? It is unable to answer the question: Benefits for whom, at whose cost? Volume 11 on Regional Economic Impacts tries to analyze this problem to some extent and, in brief, the following facts emerge:

- Hubei, the richest of the three provinces that would be affected by the project, and with the smallest population, would get the maximum share of power and flood control benefits.
- Sichuan, by far the most populous and poorest of the provinces that would be affected, would bear the brunt of the externalized cost burdens in the form of displaced

population, loss of agricultural land and what remains of its forested land. The city of Chongqing would also have to forego navigation benefits for at least 12 years or more, and later it may have to incur a large expenditure on dredging the upstream end of the reservoir.

• The project construction expenditure would be made almost entirely in Hubei province and the power generated would be used by the central China grid and the eastern grid near the industrial centre, Shanghai. Similarly, the employment generated would be largely in Hubei province, Shanghai, and the northeast region of China.

With any megaproject of this kind, disparity in distribution of costs and benefits is almost inevitable. It is unfortunate that the poorer of the three regions would have to bear the external as well as the direct cost burdens whereas Hubei province should get the maximum benefits.

Conclusion

The impression of exactitude and precision which is conveyed through equations, ratios and percentages in the CYJV study is in fact misleading, as they are based on assumptions which are unrealistic and untenable. Even a small but plausible change in some of the basic assumptions can make a dramatic difference in the present value of net benefits or in the financial rate of return. As well, the number of uncertainties with regard to costs, both internal and external to the project – for example, currency exchange rates, national economic growth rate, price inflation, and construction schedule delays – render CYJV's cost estimate an unacceptable basis for a decision which would have such enormous social and environmental ramifications.

Appendix A: Resettlement Criteria

1. The population subject to resettlement should, at a minimum, maintain its current standard of living and should have the opportunity to achieve a higher standard of living after resettlement has taken place.

2. The resettlement transition period should be minimized and adequate support of both a social and economic nature should be provided during the transition period.

3. Resettlement should achieve the social and economic reestablishment of those dislocated, on a viable productive basis, through the creation of project-funded new industrial, service sector and agricultural employment and activities. Insofar as changes in occupation are necessary, the replacement opportunities should properly recognize the social, communal, cultural, educational and vocational profile of those affected, and any changes in economic activity should be introduced on a voluntary basis.

4. The resettlement plans should have broad-based popular acceptance and the affected population should be consulted.

5. The distances the population to be relocated are required to move should be minimized and opportunities for resettling people in groups and communities should be provided.

6. The affected urban resettlements should be adjusted or replanned with adequate regard to functional structure, infrastructure provision and new site capability.

7. Houseplot and dwelling size allocation at new rural and urban sites should ensure improved conditions and take into account predictable growth of affected households.

8. The resettlement plans should minimize the loss of existing natural resources, especially agricultural lands.

9. The resettlement plans should have adequate institutional arrangements to ensure effective and timely implementation and adequate monitoring and evaluation arrangements.

10. The financial resources to carry out the relocation and development proposals should be available when and where required.

11. As a measure of mitigation in itself, the impacts of the resettlement on the natural and socio-economic environment, including cultural heritage items, should be considered acceptable.

12. Only those factories that will be economically viable should be considered for relocation and the compensation for the assets of the non-viable factories should be used to create new employment opportunities.

13. Inundated items of infrastructure such as roads, bridges, etc. should be replaced so that the previous level of service is maintained or improved.

14. Resettlement should not cause adverse socio-economic impacts on the standards of living of the host population nor on their environment.

SOURCE: CYJV, Vol. 1, pp. 17-1, 17-2

Appendix B: Probe International's Complaints to the Engineering Associations

On September 17, 1990, Probe International filed complaints against British Columbia Hydro International, Hydro-Québec International, SNC, Lavalin International, and Acres International for their work on the *Three Gorges Water Control Project Feasibility Study*. The complaints were filed with the regulatory bodies that are legally responsible for regulating the profession of engineering in the provinces of British Columbia, Quebec, and Ontario. Using the findings contained in *Damming The Three Gorges: What Dam Builders Don't Want You To Know*, Probe International accused the engineering companies of negligence, incompetence, and professional misconduct.

Probe International argued that the engineers licensed in British Columbia violated sections of their Code of Ethics as found in the August 1990 Engineers and Geoscientists Act of British Columbia, including: section 1(a) which states "He [the Engineer] will be realistic...in the preparation of all estimates, reports, statements and testimony;" section 1(b) which states "He will not distort...facts in an attempt to justify his decisions or avoid his responsibilities;" section 2 which states "The engineer will have proper regard for the safety, health and welfare of the public in the performance of his professional duties. He will regard his duty to the public safety and health as paramount;" and section 2(d) which states "He will guard against conditions which are dangerous or threatening to the environment and he will seek to ensure that all standards required by law for environmental control are met."

In Quebec, Probe International argued that the Quebec engineers violated sections of the Code of Ethics of the Ordre des ingénieurs du Québec, c. I-9, r.3., which is codified as a regulation under the Engineers Act of the Province of Quebec, including: Division II, Duties and Obligations Towards the Public, section 2.01 states that "In all aspects of his work, the engineer must respect his obligations towards man and take into account the consequences of the performance of his work on the environment and on the life, health and property of every person." Division III Duties and Obligations Towards Clients, section 3.02.04 states that "An engineer must refrain from expressing or giving contradictory or incomplete opinions or advice, and from presenting or using plans specifications and other documents which he knows to be ambiguous or which are not sufficiently explicit." Section 3.02.08 states that "The engineer shall not resort nor lend himself to nor tolerate...doubtful practices in the performance of his professional activities."

Finally, in Ontario, Probe International argued that the Ontario-licensed engineers violated their professional duties as defined in the Professional Engineers Act of Ontario, and in particular in the Code of Professional Conduct and Code of Ethics contained in the Act. Section 29(3)(a) of the Act says a member of the association may be found incompetent if "the member...has displayed in his professional responsibilities a lack of...judgment or disregard for the welfare of the public of a nature or to an extent that demonstrates the member or holder is unfit to carry out the responsibilities of a professional engineer." A member of the association may be found guilty of professional misconduct, which as defined in section 86(2) of the Regulation includes: "(a) negligence," and "(b) failure to make reasonable provision for the safeguarding of life, health or property of a person who may be affected by the work for which the practitioner is responsible." Negligence is defined under section 86(1) of the Regulation as "an act or an omission in the carrying out of the work of a practitioner that constitutes a failure to maintain the standards that a reasonable and prudent practitioner would maintain in the circumstances." Furthermore, according to the Code of Ethics contained in Regulation 538/84 of the Ontario Professional Engineers Act, under section 91.2 "A practitioner shall, (i) regard his duty to public welfare as paramount."

If, after an investigation and a public hearing, the three engineering regulatory bodies found the engineering companies guilty of violating their professional responsibilities in the course of carrying out the *Three Gorges Water Control Project Feasibility Study*, they could have taken a variety of disciplinary actions, including reprimanding, admonishing or counseling the offending member, imposing conditions on that member, levying financial fines, and suspending or revoking the engineers' licences.

In each case, the Association of Professional Engineers of Ontario, the Ordre des ingénieurs du Québec, and the Association of Professional Engineers of the Province of British Columbia — each legislated to be self-regulating bodies of the engineering profession in their own provinces — rejected Probe International's complaints of engineering negligence, incompetence, and professional misconduct. Here is a synopsis of their reasons.

Quebec Response

In Quebec, the Ordre des ingénieurs du Québec (OIQ) rejected Probe International's complaint on the grounds that "we have authority over individuals only, and none over engineering firms."* Probe International had lodged its complaint against the firms involved because CYJV had refused to reveal the names of the individual engineers that carried out the feasibility study. OIQ nonetheless carried out an internal investigation, the details of which were not disclosed to Probe International, and concluded:

> from an organizational perspective, it is impossible for us to attribute to either one or several of our Members total or even partial responsibility for any hypotheses, solutions, and recommendations. Our Code of Professional Ethics applies only to individuals, and not to committees, groups, companies, or consortia.

As for the findings in *Damming The Three Gorges: What Dam Builders Don't Want You To Know*, the OIQ dismissed them, arguing that "these criticisms represent differences in opinion among experts, differences that stem from the context of the studies, time constraints for completion of the studies, and the objectives of each study," and not from professional misconduct, negligence, or incompetence on the part of the engineering firms.

"The impossibility of charging any member of the Ordre des ingénieurs du Québec," the OIQ explained, "is primarily based on the operational structure of the project, and its terms of reference." The OIQ maintains that,

> No individual decision was taken by any particular engineer for which he/she could be held responsible. In fact, all decisions concerning this huge project were made by multidisciplinary groups following discussions, criticism, and approvals or disapprovals by internal review groups, a coordination group, then a management committee. Finally, a panel of international experts from various fields, completely independent of both the consortium and the firms named in your denunciation, made a complete review of every major

---

*July 19, 1991, response from Mr. Luc Laliberté, Syndic of the Ordre des ingénieurs du Québec, to Probe International's complaint against Hydro-Québec International, Lavalin International and SNC. Original letter in French. English translation provided by the Association of Professional Engineers of Ontario.

recommendation. China was represented by a corps of responsible individuals from every level.

"How is it possible, in this context," OIQ asks, "to attribute the responsibility for a specific act to one person in particular? Québec legislation on the practice of engineering limits me to the individual."

The OIQ acknowledged that concerned citizens all over the world wish to avoid environmental damage associated with economic development. "But, on the other hand," the OIQ explains, "given that the very essence of engineering is to take nature's forces and resources and transform them into something to improve the lot of humanity, the environment can be effected *[sic]*. Engineering's integral goals of development and environmental protection have always been, and shall always be, comprised of objectives that are difficult to reconcile. It is for this reason that the engineers must always consider the consequences of their acts upon the environment, and endeavour to minimize that impact."

That said, the OIQ concludes: "In view of the complexity of the organization, the divergent opinions of the experts, the willingness of China to proceed with energy development, the mandate awarded, and, above all, the absence of evidence for negligence and misconduct directly attributable to any individual, it is our decision to close this file."

In other words, when two or more Quebec-licensed engineers work together on a project, it is impossible to assign responsibility for any or all decisions to any of them, and therefore to guarantee the delivery of professional engineering standards consistent with those contained in Quebec's Engineers Act.

British Columbia Response

In British Columbia, the Association of Professional Engineers of the Province of British Columbia (APEBC) also carried out an investigation. As in Quebec, the details of that investigation were undisclosed to Probe International.

The APEBC decided not to proceed to a formal Inquiry into the professional conduct of professional engineers assigned to the Three Gorges Feasibility Study by B.C. Hydro International, giving the following reasons.

First, the APEBC argues, it was outside the Terms of Reference for CYJV to consider alternatives to the Three Gorges Dam.

Next, the APEBC maintains that Probe International's concerns about the environment and resettlement were "expressed as comments and unanswered questions." In any case, the APEBC goes on, "these matters were reserved for study by other agents of The Peoples [sic] Republic of China."

"The concern of Probe and its experts that the study is flawed by omissions reflects on the Terms of Reference. The study report identifies a number of sociological and environmental concerns that are related to the project, but not considered for evaluation because of the Terms of Reference." In fact, the Terms of Reference is unambiguous. Under the headline "Environment" in the Terms of Reference,* CYJV is instructed to "prepare two separate feasibility assessments, one for environmental, the second for resettlement which, inter-alia, will review, evaluate and recommend on the following subjects:"

> (a) the technical and social feasibility of plans for resettlement of inhabitants and relocation of municipalities, industry, transportation, utilities, from all project areas.
> (b) the compatibility of the resettlement and relocation plans with overall project requirements and schedules.
> (c) the adequacy of the cost estimates for compensation, resettlement and relocation, including the management of cultural property.
> (d) the adequacy of institutional arrangements for implementing the above plans on schedule, within budget, and according to specifications.
> (e) the adequacy of socio-economic impact and evaluation of the above plans, including arrangements for any ethnic minorities.
> (f) the adequacy of fisheries and water quality information in the reservoir area for the purpose of evaluating the potential of reservoir fisheries and agriculture [sic], and of approaches for linking reservoir fisheries with resettlement.
> (g) the feasibility report should review other environmental aspects such as: endangered species and habitats, health and disease, esthetics and downstream effects.

*CIPM Yangtze Joint Venture (CYJV) *Three Gorges Water Control Project Feasibility Study*, Vol. 1A, p. 1-7.

(h) review, with special consideration, the environmental problems in the Daning River Valley are to Shennongjia forest area "trade offs" *[sic]*.

It is possible, the APEBC explains, that Probe International might have been satisfied by the complementary study, which APEBC maintains included an evaluation of the socio-economic and environmental impacts. But the complementary study was canceled by the Canadian government after the Tiananmen Square massacre, leaving these issues unresolved, and leaving CYJV's recommendation to proceed with the dam unsupported.

The APEBC concludes by echoing the OIQ: "The criticisms expressed by Probe and their experts in their publication *Damming The Three Gorges*, APEBC says, "tended to be opinions, not fully supported or documented, and the discipline process of the Association is not structured to arbitrate diverse opinions."

Ontario Response

Meanwhile in Ontario, after a two-year investigation, the Association of Professional Engineers of Ontario (APEO) concluded that "there are in this case varying opinions among competent, experienced and reputable experts as to whether the Feasibility Study reflects an acceptable standard of engineering practice on the part of the CYJV in general, and Acres in particular."

Having said this, however, the APEO rejects Probe International's complaint of negligence, professional misconduct, and incompetence, on the ground that Acres International followed "generally accepted international engineering standards" as standards which "Ontario professional engineers practising outside Canada may base their work on." But the APEO fails to define those "generally accepted international engineering standards," or to identify who sets them and who enforces them, citing instead widespread support for the Feasibility Study from "reputable sources." Nor does the APEO explain how those "generally accepted international engineering standards" could deviate so dramatically from the standards of Ontario, Britain and the U.S., and the standards advocated by the U.S. Commission on Large Dams and the International Commission on Large Dams, which Probe International argued were not met by CYJV in its work on the *Three Gorges Water Control Feasibility Study*.

Probe International intends to appeal the decision by the APEO.

Endnotes

Introduction

1.   "Yao Yilin Says That for the Time Being China Will Not Consider Starting the Three Gorges Project Immediately," *Zhongguo Tongxun She*, 23 January 1989.

2.   Edward Goldsmith and Nicholas Hildyard, *The Social and Environmental Effects of Large Dams*, (San Francisco: Sierra Club Books, 1984), p. xi.

3.   CIPM Yangtze Joint Venture (CYJV), *Three Gorges Project Water Control Project Feasibility Study*, Vol. 1, p. 16-12.

4.   Personal correspondence with Vaclav Smil, 3 April 1989.

Chapter One: 1920 – 1993

1.   Robert Delfs in Peking, "China's Rivers 1: Wealth and Woe," *Far Eastern Economic Review*, 15 March 1990, p. 23.

2.   Baruch Boxer, "China's Three Gorges Dam: Questions and Prospects," *The China Quarterly* 113, March 1988, p. 99.

3.   Estimates of the cost overrun range from two to four times the original Yangtze Valley Planning Office estimate. See Xu Heshi, "Rushing to a Decision Should be Avoided; Discussion of Issues Related to the Three Gorges Project Should Follow Scientific and Democratic Principles," Tian Fang and Lin Fatang (Eds.), *Further Discussions of Macro Decision-making of the Three Gorges Project*, (Changsha: Hunan Science and Technology Press, 1989), pp. 157-161.

4.   Although Mao's 'grain-first' policy did increase the nation's grain yields, the widespread environmental degradation it triggered is undermining the nation's long-term agricultural and fisheries productivity. Land unsuited for cultivation – forests, grasslands, wetlands, and even lakes, ponds, and coastal beaches – were converted to grain fields as dictated by the central policy initiative. Deforestation, accelerated soil erosion, and desertification followed as billions of tonnes of topsoil were washed away from the land that had been stripped of its natural cover. Deforestation along the Yangtze's upper reaches combined with land reclamation efforts (the filling in of lakes and ponds) along the middle and lower reaches significantly reduced the valley's natural ability to retain floodwaters. China's traditional freshwater and coastal aquaculture suffered a sharp decline in production due to widespread land reclamation and water pollution.
      For a more detailed discussion about the environmental effects of Maoist policies in China, see Vaclav Smil, *The Bad Earth*, (New York: M.E. Sharpe, 1984).

5.   "Opinions Against Three Gorges Project Openly Published," *Zhongu Tongxun She*, 24 March 1989, [in Chinese].

6.   Economic Construction Group of the Chinese People's Political Consultative Committee, "The Three Gorges Project Should Not Go Ahead in the Short Term," *Chinese Environment and Geography* 1, no. 3, pp. 84-94.

7.   Three Gorges Project Proposal, U.S. Three Gorges Working Group, July 1985.

8.   Canadian International Development Agency (CIDA) Briefing (transcript), 14 February 1989, Ottawa.

9.   Peter G. Haines, "Canadian Competitiveness and After Sales Service", an address delivered at the Canadian Export Association's Annual Consultations with the Canadian International Development Agency, Ottawa, 10-11 June 1986.

10.   U.S. sinologist and State Department consultant, Kenneth Lieberthal, may have another explanation for why the American dam builders withdrew from the competition to conduct the Three Gorges feasibility study:

> "We must recognize that powerful interests in China will continue to oppose this dam, and they will not hesitate to campaign within the country against foreign powers that they can accuse of trying to promote construction....America should be seen as a source of prompt, friendly, high-quality, and sophisticated advice. It should avoid becoming identified as an advocate of any particular position in the domestic debate. Once the Chinese have decided to forge ahead with actual dam construction, the United States should seek to become a supplier of equipment, technical advice (including advice on project management and financing..."

Source: Kenneth Lieberthal, *United States Participation in the Three Gorges Project*, (Washington: Institute for Values in Public Policy, 1988), p. 36.

11.   Ronald Anderson, "Engineers' Beachhead in Third World Could Be Boon to Canadian Industry," *The Globe and Mail* (Toronto), 10 November 1987.

12.   CIPM Yangtze Joint Venture (CYJV), *Three Gorges Project Water Control Project Feasibility Study*, Vol. 1A, p. 1-2.

13.   CIDA signed a number of contracts in the autumn of 1988 with consultants across the country to fulfil this monitoring function. The firms included Both, Belle, and Rob of Montreal, Quebec, Canergie Inc. of Baie d'Urfe, Quebec, Benjamin Development Planning Consultants of Calgary, Alberta, and R.L. Walker and Partners of Ottawa, Ontario. All contract documents were secured by Probe International using the Access to Information Law.

14.   Contract of October 14, 1988, between the Canadian Department of Supply and Services and Peter Haines for $208,677.60. This contract was also secured by Probe International using the Access to Information Law. For his part in promoting the export of Canadian engineering expertise, Peter Haines was awarded in 1987 the highest honour by the Association of Professional Engineers of Ontario (APEO) – the Professional Engineers Gold Medal. In announcing the award, the APEO explained that Peter Haines' department within CIDA, "is involved in 1,234 projects valued at about $7.4 billion." The APEO went on, "It is largely because of Engineer Haines' personal efforts and hard work that Canadian engineering companies have won these projects [citing the Three Gorges Project] in the face of severe international competition." See APEO announcement, "In Recognition of Excellence" 1987. Also see Anderson, "Engineers' Beachhead," *The Globe and Mail*, 10 November 1987.

15.   CIDA briefing (transcript), 14 February 1989.

16.   In China, the Yangtze River is known as Changjiang. See, Dai Qing, *Changjiang Changjiang*, (Guizhou People's Press, 1989), [in Chinese].

17.   "Vigorous Debate Delays China Dam," Christian Science Monitor, 1 March 1989, p. 6.

18.   Ibid.

19.   Frederic A. Moritz, "China's Politics of the Environment," *Christian Science Monitor*, 24 August 1989, p. 18.

20.   Jan Wong, "Faced death penalty as dissident, Chinese woman leaves for U.S.," *The Globe and Mail*, 23 December 1991.

21.   Jan Wong, "U.S. wins no concessions in Beijing," *The Globe and Mail*, 18 November 1991.

22.   Personal correspondence with Douglas Lindores, Senior Vice-President, CIDA, 1 November 1991.

23.   "Body Set Up to Study Huge China Dam, Many Scientists in Favour," *Reuter News Service*, 15 July 1990.

24.   Willy Wo-lap Lam, "Decision expected on Three Gorges dam plan," *South China Morning Post*, 6 July 1990.

25.   John Fox, "Chinese dam study flawed," *The Financial Post*, 20 September 1990. "Big Yangtze Dam Plan Challenged in Canada," *International Herald Tribune*, 20 September 1990. "Environmentalists Demand Review of Canadian Engineering Report," *Engineering Dimensions*, November/December 1990. Ellen Saenger, "Dammed if they do," *British Columbia Report*, 8 October 1990. Moira Farrow, "China dam study under fire," *Vancouver Sun*, 19 September 1990.

26.   Geoffrey Crothall, "China: Backing For Gorges Project," *South China Morning Post*, 6 April 1991.

27.   Ibid.

28.   James L. Tyson, "Critics Urge China to Consult on Dam Plan," *The Cristian Science Monitor*, 22 July 1991.

29.   Graham Hutchings, "Peking To Approve Flooding of Farms By Hydroelectric Project," *The Daily Telegraph*, 10 October 1991.

30.   "China: Watering Down Flood Prevention," *South China Morning Post*, 7 August 1991.

31.   Willy Wo-lap Lam, "Journal Attacks Gorges Hydro-Electric Scheme," *South China Morning Post*, 6 January 1992.

32.   Geoffrey Crothall, "Academics Denounce Three Gorges Report," *South China Morning Post*, 23 March 1992.

33.   Personal correspondence with Michael Wilson, Canada's International Trade Minister, 27 January 1993.

34.   "China Wants Foreigners To Help It Build Dam," *Associated Press, Reuter News Service*, 19 February 1992.

35.   Daniel Kwan and Geoffrey Crothall, "One Third of Delegates Refuse to Approve The Three Gorges Project," *South China Morning Post*, 4 April 1992.

36.   As reported in "China Passes Dam Project, Many Abstentions," *Reuter News Service*, 3 April 1992.

37.   Linda Hossie, "CIDA strikes blow at big-dam agenda," *The Globe and Mail*, 6 April 1992.

38.   Ibid.

39.   Yosahiko Lakurai, "China Lets Foreigners Enter Finance Market," *Nikkei Weekly*, 29 August 1992.

40.   "Construction Bank Establishes Branch To Fund Three Gorges Project," *Xinhua News Agency* as reported by *BBC Monitoring Service*, 14 October 1992.

41.   James McGregor, "China Bulldozing Ahead on Dam Project," *The Wall Street Journal*, 19 January 1993.

42.   Pan Jiazheng and Zhang Jinsheng, "Hydropower development in China," *International Water Power & Dam Construction*, February 1993.

Chapter Three: Resettlement Plans

1.   Tian Fang and Lin Fatang, "Population Resettlement and Economic Development in the Three Gorges Reservoir Area," *Chinese Geography and Environment* 1, no. 4, 1988, pp. 90-100.

2.   CIPM Yangtze Joint Venture (CYJV), *Three Gorges Water Control Project Feasibility Study*, Vol. 1, p. 1-4.

3.   CYJV, Vol. 1, p. 1-3.

4.   CYJV, Vol. 1, p. 16-8.

5.   CYJV, Vol. 9, p. 2-2.

6.   CYJV, Vol. 9, p. 2-9.

7.   CYJV, Vol. 1A, p. 5-1.

8.   The Yangtze catchment is the site of a dramatic surge in forest loss. See Vaclav Smil, "Deforestation in China," *Ambio* 12, no. 5, 1983, pp. 226-231.

9.   CYJV, Vol. 8A, p. 5-6.

10.   The NPL plus two metres is used to account for wind-produced wave action along the reservoir shoreline. CYJV states that two metres may be an insufficient safety margin, particularly in the event of unstable slopes and extraordinary wave action which could result in costly damage to urban areas. In what appears to be a concession to the Chinese proponents, CYJV states it "has accepted the 2 m [metre] concept for the purposes of this study, but allowance has also been made in the project construction cost estimates to deal with any future unstable conditions along reservoir requisition lines" (CYJV, Vol. 9, p. 5-4).

11.   CYJV, Vol. 8A, p. 3-22.

12.   CYJV, Vol. 8A, p. 5-9.

13.   CYJV, Vol. 9, p. 2-9.

14.   Fang Zongdai and Wang Shouzhong, "Resettlement Problem of the Three Gorges Project," *Chinese Geography and Environment* 1, no. 4, 1988, p. 86.

15.   CYJV, Vol. 8A, p. 5-3.

16.   CYJV, Vol. 9, p. 5-16.

17. The World Bank-funded Carajás iron project in Brazil, completed in 1984, is one example. Since January 1988, the mine, railway and port facilities built for export of unprocessed iron ore (after Bank disbursements were completed) have also been used as part of a scheme to produce pig iron in smelters along the railway route. The charcoal used by the smelters has tremendous potential for speeding deforestation in the eastern Amazon. (See P.M. Fearnside, "The Charcoal of Carajás: Pig-iron Smelting Threatens the Forests of Brazil's Eastern Region," *Ambio* 18, no. 2, 1989.) In this case, the potential threat of the pig-iron plan had been known before the railway was built but the World Bank chose not to consider these plans, which were administratively separate from the bank-financed project. (See P.M. Fearnside and J.M. Rankin, "Jari and Carajás: The Uncertain Future of Large Silvicultural Plantations in the Amazon," *Interciencia* 7, no. 6, 1982, p. 326.) Once smelting began and the danger was apparent to the world, the Bank had no leverage with which to induce compliance with clauses in the loan agreement committing Brazil to protect the environment along the railway route.

18. CYJV, Vol. 9, p. 3-4.

19. Yao Ianguo, "Three Gorges Project: Dream and Reality," *Beijing Review* 32, no. 27, pp. 19-30.

20. CYJV, Vol. 1A, p. 2-10.

21. CYJV, Vol. 9, p. 7-7.

22. See D.M. Lampton, "Water Politics," *The China Business Review*, July-August 1983, pp. 10-17; Liu Chiangming and L.J.C. Ma, "Interbasin Water Transfer in China," *Geographical Review* 73, no. 3, 1983, pp. 253-270; and CYJV, Vol. 8A, p. 6-14.

23. CYJV, Vol. 9, p. 7-7.

24. Ibid.

25. Lampton, "Water Politics," p. 16.

26. P.M. Fearnside, "China's Three Gorges Dam: 'Fatal' Project or Step Toward Modernization?" *World Development* 16, no. 5, 1988, pp. 615-630.

27. CYJV, Vol. 9, p. 3-2.

28. Tian and Lin, "Population Resettlement," p. 97.

29. CYJV, Vol. 9, p. 3-3.

30. CYJV, Vol. 9, p. 4-21.

31. CYJV, Vol. 9, p. 7-9.

32. Tian and Lin, "Population Resettlement," p. 93.

33. CYJV, Vol. 9, p. 7-19.

34. CYJV, Vol. 9, p. 7-18.

35. To compare flood risk with acceptable frost risk, citrus is presently grown in areas where frost events kill, at most, some trees once every 25 years. See CYJV, Vol. 9, p. 4-15.

36. CYJV, Vol. 9, p. 2-4.

37.   See R. Gore and B. Dale, "Journey to China's Far West," *National Geographic* 157, no. 3, 1980, p. 321; A. Samagalski and M. Buckley, *China: A Travel Survival Kit*, (South Yara, Victoria, Australia: Lonely Planet Publications, 1980), p. 57; and Fearnside, "China's Three Gorges Dam," p. 619.

38.   R.J.A. Goodland, *Tribal Peoples and Economic Development: Human Ecological Considerations*, (Washington, D.C.: The World Bank, 1982).

39.   CYJV, Vol. 9, p. 4-10.

40.   Personal communication, R.J.A. Goodland, 1987.

41.   CYJV, Vol. 9, p. 5-7.

42.   CYJV, Vol. 9, p. 5-10.

43.   Ibid.

44.   *Document No. 12 (1986) of the Central Committee of the Communist Party*, Vol. 9, Annex 1, p. 2.

45.   CYJV, Vol. 9, p. 4-26.

46.   CYJV, Vol. 9, Annex 1, p. 2.

47.   CYJV, Vol. 9, p. 2-4.

48.   CYJV, Vol. 1A, p. 2-14.

49.   CYJV, Vol. 1, p. 17-17.

50.   CYJV, Vol. 9, p. 7-11.

51.   Ibid.

52.   CYJV, Vol. 9, p. 7-5.

53.   CYJV, Vol. 1, p. 17-1.

54.   Even in Canada, where the CYJV feasibility study was released in compliance with Probe International's request using Canada's Access to Information Law, information was omitted from the report, as indicated by a stamp specifying, for example, "information deleted pursuant to 19(1), Access to Information Act," in Volume 9 (Resettlement), p. 5-16.

55.   CYJV, Vol. 8A, p. 6-14.

Chapter Four: Reservoir: Environmental Impacts

1.   CIPM Yangtze Joint Venture (CYJV), *Three Gorges Water Control Project Feasibility Study*, Vol. 8A, p. 1-4.

2.   CYJV, Vol. 8B, p. 2-4.

3.   CYJV, Vol. 8, pp. 6-87, 6-88.

4.   CYJV, Vol. 1, p. 16-10.

Chapter Six: Dowstream Environmental Impacts

1.   CIPM Yangtze Joint Venture (CYJV), *Three Gorges Water Control Project Feasibility Study*, Vol. 1, p. 16-12.

2.   Lacking specific data, CYJV made certain assumptions and calculations as follows: an increase of 0.8 metres in the Yangtze River level during the dry season would result in little change in the level and surface area of Dongting Lake; conversely, for May and October, a decrease of 0.7 metres and 1.3 metres, respectively, would significantly affect surface areas.   For Poyang Lake, CYJV suggests that an increase of 0.5 metres in the Yangtze River level during the dry season would have little effect, but that the same increase from mid-March to mid-April would result in flooding of an additional 200 square kilometres.   To analyze the actual impact of the dam on river and lake levels, the development of stage-area curves is required.   YVPO did not provide these to CYJV; CYJV did draft one for Poyang Lake but used data from an unidentified source.

   CYJV recommends that the Chinese establish a computerized hydrologic database for the downstream components of the river.   This would make it possible to test the effect of different water release patterns, thereby avoiding costly adverse impacts.   Whether the cost of conducting this analysis is included in CYJV's cost-benefit calculations is not clear.

3.   CYJV, Vol. 8, p. 6-94.

Chapter Seven: Unresolved Issues: Perspective from China

1.   See Chinese Academy of Sciences, Leading Group of the Three Gorges Project Ecology and Environment Research Project, *Collected Papers on Ecological and Environmental Impact of the Three Gorges Project and Countermeasures*, (Beijing: Science Press, 1987), [in Chinese], and *Ecological and Environmental Impact of the Three Gorges Project and Countermeasures*, (Beijing: Science Press, 1988), [in Chinese].   This research was conducted by the State Science and Technology Commission from 1983 to 1986 in cooperation with the Chinese Academy of Sciences.

2.   Tian Fang and Lin Fatang, *Discussions of Macro Decision-making of the Three Gorges Project*, (Changsha: Hunan Science and Technology Press, 1988), [in Chinese], and *Further Discussions of Macro Decision-making of the Three Gorges Project*, (Changsha: Hunan Science and Technology Press, 1989), [in Chinese].   See also, Dai Qing, *Changjiang, Changjiang*, (Guizhou People's Press, 1989), [in Chinese].

3.   "China Plans New Resettlement Rules," *Water Power and Dam Construction*, March 1990, p. 2.

4.   In a recent review of Chinese literature, Luk and Whitney noted that with few exceptions, virtually none of the authors opposed to the TGP cited population displacement as a major argument to support their opposition to the project.   See S.H. Luk and J. Whitney, "Introduction to the Special Issue on the Three Gorges Project, Part II," *Chinese Geography and Environment* 1, no. 4, 1988, p. 14.

5.   Tian Fang and Lin Fatang, "Population Resettlement and Economic Development in the Three Gorges Reservoir Area," *Chinese Geography and Environment* 1, no. 4, pp. 90-100.

6.   Chen Guojie, "The TGP Area is Already Overpopulated and Local Resettlement Will Be a Disaster," *Soil and Water Conservation Bulletin* 7, no. 5, p. 42, [in Chinese].

7.   Shi Deming, Yang Yangsheng, and Lu Xixi, "The Impact of Soil Erosion on Sediment Sources in the TGP Reservoir Area and Countermeasures," Chinese Academy

of Sciences, *Collected Papers on Ecological and Environmental Impact of the Three Gorges Project and Countermeasures*, pp. 498-521.

8.   Tian and Lin, "Population Resettlement and Economic Development," pp. 90-100.

9.   Chen, "The TGP Area is Already Overpopulated," pp. 42-46.

10.   Fang Zongdai and Wang Shouzhong, "Resettlement Problem of the Three Gorges Project," *Chinese Geography and Environment* 1, no. 3, 1988, pp. 81-89.

11.   Shi, Yang, and Lu, "The Impact of Soil Erosion," pp. 498-521; Wang Chaojun "Comprehensive Assessment of the Ecological and Environmental Impact of the Three Gorges Project," *Chinese Geography and Environment* 1, no. 3, 1988, pp. 45-83; Shi Deming "Speeding Up the Control of Soil Erosion in the Upper Reaches of the Changjiang River is a Fundamental Requirement for the Construction of the TGP," Tian and Lin, *Further Discussions*, pp. 326-329; Fang Zongdai, "Sediment Management in the Key to the Success or Failure of Water Conservancy Projects and There is a Need to Properly Evaluate the Impact of Human Activities on the Increase of Sediment Production," ibid., pp. 282-291; and He Naiwei, "Downstream Effects of Soil and Water Conservation and Protection Forestry in the Upper Reaches of the Changjiang River," ibid., pp. 317-325.

12.   CYJV looked at the annual sediment loads of the Yangtze at Yichang and predicted no obvious trends, either increases or decreases. The lack of a dramatic rise in sediment load can be attributed in part to much of the sediment being trapped behind dams upstream. See Chinese Academy of Sciences, *Ecological and Environmental Impact*, p. 163.

13.   Once these reservoirs become clogged with sediment, which could be in 50 years or less, more sediment will be flushed through the reservoirs and transported by the Yangtze, causing a large increase in the sediment load at Three Gorges; see Gu Hengyu, et al., "Sediment Sources and Trends of Sedimentation in the TGP Reservoir Area," Chinese Academy of Sciences, *Collected Papers*, pp. 522-541.

14.   CIPM Yangtze Joint Venture (CYJV), *Three Gorges Water Control Project Feasibility Study*, Vol. 1, p. 11-14.

15.   Qian Ning, Zhang Ren, and Chen Zhicong, "Some Aspects of Sedimentation at the Three Gorges Project," *Chinese Geography and Environment* 1, no. 4, 1988, pp. 26-65.

16.   Because of massive sedimentation upstream of the Sanmen Gorge Dam, the structure had to be reconstructed in the early 1970s. Zeng Qinghua, a former senior official with the Hydropower Research Institute, claims that the operating principle of "storing clear water and discharging the muddy water" (as advocated for the TGP) has been quite successful in flushing out sediment that has built up behind the dam. However, this operating procedure is proving ineffective in removing sediment 120 kilometres upstream. See Zeng Qinghua, "Experience from the Sanmenxia Reservoir Indicates that the Operating Mode of 'Storing the Clear Water and Discharging the Turbid' Does Not Solve the Problem of Sedimentation at the Backwater Zone of the TGP Reservoir," Tian and Lin, *Further Discussions*, pp. 330-333.

17.   Sun Yueqi, 90, is leader of the Chinese People's Political Consultative Committee (CPPCC) and is one of China's most respected and dedicated detractors of the TGP. Fang is a former senior engineer with the Ministry of Water Resources. See Fang Zongdai, "The Flood Prevention Function of the Three Gorges Project – Disadvantages Outweigh Advantages," *Chinese Geography and Environment* 1, no. 4, pp. 66-80; Fang, "Sediment Management," pp. 282-291; and Sun Yueqi, "Basic Principles for the Changjiang River Basin Planning," Tian and Lin, *Further Discussions*, pp. 124-134.

18.   Wang, "Comprehensive Assessment," pp. 45-83; and, Qian, Zhang, and Chen, "Some Aspects of Sedimentation," pp. 26-65.

Other references for chapter seven

Economic Construction Group of the Chinese People's Political Consultative Committee. "The Three Gorges Project Should Not Go Ahead in the Short Term." *Chinese Environment and Geography* 1, no. 3, 1988, pp. 84-94.

Fearnside, P.M. "China's Three Gorges Dam: 'Fatal' Project or Step Toward Modernization?" *World Development* 16, no. 5, 1988, pp. 615-630.

Luk, S.H. and Whitney, J. "Introduction to the Special Issue on The Three Gorges Project, Part I." *Chinese Geography and Environment* 1, no. 3, 1988, pp. 3-13.

Xu Heshi. "Rushing to a Decision Should be Avoided; Discussion of Issues Related to the TGP Should Follow Scientific and Democratic Principles." Tian Fang and Lin Fatang (Eds.) *Further Discussions of Macro Decision-making of the Three Gorges Project.* Changsha: Hunan Science and Technology Press, 1989, in Chinese.

Chapter Eight: Flood Control Analysis

1.   CIPM Yangtze Joint Venture (CYJV), *Three Gorges Water Control Project Feasibility Study*, Vol. 4, p. 1.

2.   CYJV, Vol. 7, p. 1-21, and p. 9-4, Table 9.2.

3.   CYJV, Vol. 7, p. 7-1.

4.   CYJV, Vol. 7, p. 7-10.

5.   CYJV, Vol. 1, p. 11-11.

6.   CYJV, Vol. 7, p. 4-3.

7.   CYJV, Vol. 7, p. 1-2.

8.   CYJV, Vol. 7, p. 9-32.

9.   CYJV, Vol. 7G, p. 1-2.

10.   CYJV, Vol. 7F, p. 6-1.

11.   CYJV, Vol. 1, p. 16-3.

Chapter Nine: Missing Energy Perspectives

1.   Baruch Boxer, "China's Three Gorges Dam: Question and Prospects," *The China Quarterly* 113, 1988, pp. 94-108.

2.   Actual kilowatt-hours per capita figures for 1986 to 1987 were: East China 370, Central China 250, Eastern Sichuan 200, Brazil 1400, Mexico 1200, U.S.A. 10,700, and Canada 18,700. See UNO, *Yearbook of Energy Statistics*.

3.   V. Smil, "China's Energy," (1990). Report prepared for the Office of Technology Assessment of the United States Congress.

4.   See State Statistical Bureau (SSB), *China Statistical Yearbook 1988*, (Beijing, 1988); and V. Smil, *Energy in China's Modernization*, (Armonk, New York: M.E. Sharpe, 1988).

5. See SSB, *China Statistical Yearbook 1988.*

6. See Sun Yueqi, "Why I Am Against the Project," *Beijing Review* 32, no. 27, 1989, pp. 31-35; and Tian, Fang and Lin Fatang, (Eds.), *More on the General Decision Concerning the Three Gorges Project*, (Changsha: Hunan Provincial Science and Technology Publishing House, 1989), [in Chinese].

7. SSB, *China Statistical Yearbook 1988.*

Other references for chapter nine

Education and Science Society. 1986. *Symposium on the Three Gorges Project.* Education and Science Society, New York.

Fearnside, Philip M. 1988. "China's Three Gorges Dam: 'Fatal' project or step toward modernization?" *World Development* 16, pp. 615-630.

Smil, Vaclav. 1990. *General Energetics*, John Wiley, New York.

Chapter Ten: Dam Safety Analysis

1. The methodology used for estimating ground acceleration described in International Commission on Large Dams (ICOLD) bulletin 46 is based on U.S. Bureau of Reclamation procedures. Within the U.S. these procedures have been criticized by engineering geologists as too optimistic in estimating MCEs. For this MCE, a horizontal ground acceleration of [0.17g] – 0.17 multiplied by acceleration due to gravity (g) – was estimated by CYJV as a "starting point for safety evaluation." Also, CYJV underestimates the vertical acceleration as [0.1g], two-thirds that of the horizontal acceleration value.

2. CIPM Yangtze Joint Venture (CYJV), *Three Gorges Water Control Project Feasibility Study*, Vol. 4, p. 4-14.

3. "Earthquake forces considered in the standard method for design of concrete gravity dams are seriously deficient and this has led to a number of bad features in the design practice....Because the dynamic nature of the earthquake response problem is ignored, the specifications for earthquake forces in the standard design loads for concrete gravity dams is seriously deficient." Source: A.K. Chopra, "An Evaluation of the Koyna Dam Earthquake," presented at the International Commission on Large Dams (ICOLD) Conference Proceedings, University of Southampton, 1976.

4. CYJV, Vol. 4, p. 5-11.

5. CYJV, Vol. 4, p. 12-7.

6. CYJV, Vol. 4, p. 12-12.

Chapter Eleven: Sedimentation Analysis

1. CIPM Yangtze Joint Venture (CYJV), *Three Gorges Water Control Project Feasibility Study*, Vol. 1, p. 11-14.

2. CYJV, Vol. 5, p. 1-6.

3. CYJV, Vol. 5, p. 1-4.

4. CYJV, Vol. 5C, fig. 5.3.

5.   CYJV, Vol. 5, pp. 6-16, 6-30.

6.   CYJV, Vol. 5, p. 2-4.

7.   CYJV, Vol. 5, p. 8-3.

8.   CYJV, Vol. 1, p. 11-2.

9.   1. Empirical regime equations based on slope data from other irrigation canals and natural rivers.
     2. Predictions of slope based on sediment transport equations and what is assumed to be the characteristics of the bed load material.
     3. A one-dimensional numerical model that simulates the flow of sediment through the reservoir and calculates deposition and scour from successive floods based on the hydrodynamic and sediment transport equations.

10.   American Society of Civil Engineers, Sedimentation Manual, p. 128.

11.   "According to the Yichang Hydrological Gauging Stations's 30 year survey, the Yangtze river's suspended load averages 526×10^6t annually and the bed load is 8.6×10^6t." Pan Jiazheng and Zhang Jinsheng, "The Three Gorges project goes ahead in China," *International Water Power and Dam Construction*, February 1993.

12.   CYJV, Vol. 5, p. 2-3.

13.   Ibid.

14.   CYJV, Vol. 4, p. 3-16.

15.   CYJV, Vol. 5D, fig. 3.1.

16.   CYJV, Vol. 5J, p. 4-7.

Chapter Twelve: Economic and Financial Aspects

1.   CIPM Yangtze Joint Venture (CYJV), *Three Gorges Water Control Project Feasibility Study*, Vol. 1, p. 1-5.

2.   CYJV, Vol. 1, p. 2-1.

3.   CYJV, Vol. 3, p. 1-3.

4.   CYJV, Vol. 1, p. 1-5.

5.   CYJV, Vol. 3, p. 5-14.

6.   See Vijay Paranjpye, *High Dams on the Narmada*, (New Delhi: Indian National Trust for Art and Cultural Heritage, 1990).

7.   CYJV, Vol. 3, p. 7-6.

8.   CYJV, Vol. 3, p. 3-8.

9.   CYJV, Vol. 3, p. 7-9.

10.   CYJV, Vol. 3, p. 7-9.

11.   CYJV, Vol. 3, pp. 4-2, 4-3.

12.   CYJV, Vol. 3, p. 4-10.

13.   Ibid.

14.   CYJV, Vol. 1, pp. 2-2, 17-3, 17-19; Vol. 3, pp. 1-2, 1-4.

INDEX

A

Access to Information law (Canadian) xxi, 11, 12, 77, 153
Acres International 9-10, 15-16, 19
agriculture xx, 1, 6, 23-25, 40, 48, 50-51, 68, 76, 83-84, 102, 126
  land submerged 23, 33, 37, 40, 154, 157, 160
  replacement land (*see also* compensation) 23, 48-49, 50-51, 91
alternatives 30, 147
  energy 121-123, 125
  flood control 113-114
  navigation 31, 97
American Consulting Engineers Council (*see* U.S. Three Gorges Working Group)
aquaculture (*see* fisheries, aquaculture)
Asian Development Bank 8, 22
Aswan High Dam 34, 125
Auburn Dam 131

B

Balbina Dam 57
Bechtel Civil and Mineral, Inc. (*see* U.S. Three Gorges Working Group)
Blackwelder, Brent xxi
Boxer, Baruch 3
British Columbia Hydro 10, 15-16, 19

C

Canada (*see also* Canadian International Development Agency) 8, 10, 12, 14, 15, 119, 124
  mercury experience 70-73
Canadian International Development Agency xxi, 9-10, 11-12, 14, 19, 21, 34, 36, 37
Central Committee of the Communist Party 7, 55
Chinese People's Political Consultative Committee 7-8, 13, 17-18, 96
Chongqing 6, 21, 24, 28, 29, 31, 36-37, 41, 42, 45, 46, 95, 96, 97, 99, 107, 111, 112, 139, 140, 144, 154, 155, 157, 160
CIDA (*see* Canadian International Development Agency)
coast 25, 84-86, 133
  erosion 112, 144
cofferdams 31, 127, 131
Colorado River 62
Columbia River 9, 62
compensation (*see also* land, replacement) 23-24, 26, 37, 38, 39-40, 41, 42, 44, 52, 83, 156
  exclusion from 52-53, 157
complementary studies 11-12, 15
consultation 24, 55-57, 90
Coopers and Lybrand (*see* U.S. Three Gorges Working Group)
costs (*see also* financing) 4, 7, 26, 33, 57, 88, 99, 111-112
  analysis 145-160
  decommissioning 132
  design requirements 128, 131, 154
  environmental mitigation 80, 153-155
  erosion mitigation 78, 93, 112, 131, 144, 154
  flood damage 111-112, 114-115, 148-149
  large plants 124
  navigation 98, 155
  overruns 32, 44, 152-153
  power 149-150, 153
  resettlement 32, 44, 53-54, 91-92, 99, 111, 150, 153-155, 156-157
  sedimentation 31, 95, 97, 125, 135, 144, 154, 160
  total 33, 145, 155, 157, 159
CPPCC (*see* Chinese People's Political Consultative Committee)
crane 82
  Siberian 25, 82
Cree 71
Cultural Revolution 3

State Council 6, 7, 11, 13, 45, 89
State Planning Commission 11, 89
Stone and Webster Engineering
    Corp. (*see* U.S. Three Gorges
    Working Group)
Sun Yat-sen 2

T

Tarbela Dam 108, 130, 131
terms of reference 10, 11, 36
Three Gorges Working Group (*see*
    U.S. Three Gorges Working
    Group)
Tian Fang 46-47, 89, 91
Tiananmen Square xx, xxiii, 14-15,
    17, 19
Tibet 1, 51
tributaries 25, 59, 65, 68, 86, 113,
    146
    dams 4, 5, 83, 85, 122
    flooding 17, 24, 25, 42, 68, 76,
        100, 102

U

United States 3, 8, 9, 15, 22, 119
    critical of dam 5
United States Army Corps of
    Engineers (*see also* U.S.
    Three Gorges Working
    Group) 8, 19
United States Bureau of Reclama-
    tion (*see also* U.S. Three
    Gorges Working Group) 3, 8,
    19
U.S. Three Gorges Working Group
    8, 10

V

Vaiont Dam 129

W

Wanxian 33, 66
war 3, 5, 17-18, 126
water quality (*see* pollution)
wetlands 25, 86
wildlife 68-69, 76, 85, 86

endangered 6, 25, 62, 68, 80-81
World Bank xviii, xxii, 8, 10, 15, 21,
    22, 23, 32, 36, 43, 47, 52, 90,
    150-151, 158
Wuhan 21, 30, 74, 76, 83, 100, 103,
    114

Y

Yangtze River
    flooding 1-2, 3, 17, 26, 37-39, 76,
        77, 102
    river basin 1, 63, 76-78, 80, 81,
        146-147
Yangtze Valley Planning Office 3, 4,
    5, 6, 8, 16, 27, 38, 39, 40, 41,
    42, 43, 48, 52, 93, 100, 136,
    137, 138, 156
Yangtze! Yangtze! 13, 14
Yao Yilin xx, 13
Yellow River 4, 28, 34, 62, 89, 95,
    135
Yichang 21, 22, 31, 74, 96, 135
Yichang Hydrological Gauging
    Station 27, 141
YVPO (*see* Yangtze Valley Planning
    Office)